The Water
We Drink

The Water
We Drink

Water Quality and Its
Effects on Health

Joshua I. Barzilay, M.D.
Winkler G. Weinberg, M.D.
J. William Eley, M.D., M.P.H.

Rutgers University Press

New Brunswick, New Jersey, and London

Library of Congress Cataloging-in-Publication Data
Barzilay, Joshua I.
  The water we drink : water quality and its effects on health /
Joshua I. Barzilay, Winkler G. Weinberg, J. William Eley.
    p.   cm.
  Includes bibliographical references and index.
  ISBN 0-8135-2672-8 (cloth : alk. paper). — ISBN 0-8135-2673-6
(pbk. : alk. paper)
  1. Drinking water—Health aspects. I. Weinberg, Winkler G.,
1952– . II. Eley, J. William, 1957– . III. Title.
RA591.B38   1999
613.2'87—dc21   98-55318
                                                                CIP

British Cataloging-in-Publication Data for this book is available from
the British Library

Contents

Part Three
Drinking Water and the Consumer

The Water
We Drink

Introduction

Why write a book about drinking water? In an era of consumer demand for quality products and for information regarding those products, it seems appropriate to have a guidebook regarding the "commodity" called drinking water. Interestingly, few such popular books exist. This is surprising considering the ubiquity of water and its use in all facets of life. On the other hand, it is perhaps because of the ready availability of high-quality drinking water in this country and the Western world that few people bother to inquire about it. Yet drinking water is a commodity like any other, with its own issues of processing, quality control, safety, and risk. What is more, its use affects all of us. Therefore, a book regarding drinking water should be of interest to all.

Another reason for writing a book about drinking water is that it has major effects on our health. There are very few things in life that are so common to mankind as drinking water. Fifty percent to 60 percent of our body weight consists of water. We all drink water and water-based fluids daily to replenish ourselves. What is not well appreciated is the fact that any small amounts of contaminants found in drinking water—as small as they might be—may, over a lifetime, have a cumulative and deleterious effect on health. This is especially true of factors that underlie many of the chronic illnesses that are becoming increasingly prevalent in our society as the population ages. Small changes in any of these contaminants may have profound long-term effects on individual and societal health. What is more, pollution of one drinking water source is ultimately connected to drinking water sources elsewhere, and the effect is magnified many times. Enhancing awareness of these facts is another purpose of this book.

Finally, the production of drinking water is a large enterprise. It is an enterprise that includes companies vying for water rights, companies that process and treat water, and companies that bottle water. Numerous government agencies oversee the industry that produces drinking water so as to maintain high standards of quality. A significant portion of tax revenues is expended on maintaining this infrastructure. These facts, however, are underappreciated or unknown. When we turn on the tap, we expect to have high-quality drinking water. Almost all of us do not consider how the technology that enables us to do this came to be or appreciate the fact that this is a relatively new development in mankind's history. It is another purpose of this book to make the reader aware of these points.

This book is written from a medical perspective. Its aim is to inform the reader of health issues that relate to drinking water quality. Most of the issues that we deal with have arisen only recently and reflect either the effects of widespread use of chemicals or the aging of the population.

Among the issues that we deal with are the effects of drinking water's mineral content on cardiovascular and bone health; the role of drinking water's metal content on the nervous system, especially with regard to Alzheimer's disease; and the increased risks of cancer and infertility owing to industrial chemicals that pollute drinking water supplies. In addition, we discuss drinking water as it relates to newer infectious diseases that affect people with compromised immune systems.

A note about the organization of this book. In order to put the issue of drinking water quality and its effects on health into perspective, we first review the history of water, disease, and sanitation. Then we describe the manner in which drinking water is regulated today, including a chapter on water sources and water treatment. The central portion of the book deals with health issues. At the conclusion of the book are chapters regarding bottled water and methods of water purification. We have included appendices listing water contaminants and other dissolved elements.

This book is intended for educated consumers. We cover a great deal of information in these pages, and not all chapters will be of equal interest to every individual reader. We encourage the reader to explore those chapters that interest him or her. Most chapters are self-contained and can be read separately. We hope that the reader will find this book both pleasurable and informative. We also hope that upon completion of this book the reader will appreciate that the issues involving drinking water quality and illness are very much a reflection of the economic, social, and health issues facing our society today.

Part One

General

Information

1

A Brief History of Drinking Water, Sanitation, and Disease

The concept that there is a connection between water quality, sanitation, and the presence of disease was known in biblical and classical times. Efforts at safekeeping water purity, maintaining access to waters of high quality, and providing sewage disposal were widely practiced. With the demise of the Roman Empire and the beginning of the Middle Ages, these precepts were largely forgotten, and infectious illnesses secondary to polluted waters became commonplace. Only with the ascendancy of the scientific method and the discoveries of microbiology in the last one hundred years has the connection between water quality, sanitation, and health once again been discovered.

In more recent times, interest in water quality has shifted from an emphasis on infectious diseases to an emphasis on chemical pollution. This is a direct result of the industrialization of Western societies. As will become clear, this pollution is no less threatening to the health and welfare of the community than were the water-borne outbreaks of infectious diseases in the past.

In the present chapter we briefly review the history of drinking water, sanitation, and disease so as to give the reader a perspective of the evolution of water quality and safety. This will allow the reader to better appreciate the efforts being made today to keep our waters clean.

Biblical and Classical Times

In the ancient world there was a general awareness of the need for sanitation and for water that was safe for consumption. In the Bible there are several references to water cleanliness. In Genesis 29:8 we read that Jacob, upon meeting the shepherdess Rachel, had to remove a stone from the opening of a well so as to allow her to water her flock. In all likelihood the stone had been used to cover the well to keep the well waters cool and clean. In II Kings 2:19–22 one of the first instances of water purification is recounted. The people of Jericho approached the prophet Elisha, complaining that their waters were unfit for drinking or use in farming. The prophet ordered them to bring a bowl of "salt," which he then poured into a spring. The waters were then "healed" and usable. Although the meaning of the term *salt* is not clear, this biblical description of "healing" the waters conveys the sense of disinfection or treatment. Later the prophet Ezekiel (47:9) also mentions the "healing" of polluted water and its salutary effect. The greatest contribution to water safety in the Old Testament period, however, came from public sanitation. In Leviticus and Deuteronomy basic methods of public sanitation are enumerated. Chief among these was the disposal of human waste away from areas of human habitation (see Deuteronomy 23:12–13).

Later, in the Talmudic and early Christian era (200 B.C.E. to 400 C.E.), further public sanitation laws were enacted to protect water quality. In midwinter, as we read in the Babylonian Talmud, the judicial authorities in Palestine sent out officials for the purpose of digging wells and repairing damaged reservoirs and aqueducts (Tractate Shekalim 1:1). If water was not available locally, aqueducts were built. Cisterns and wells had to be kept clean, and no waste could be thrown into them. It was forbidden to have a cemetery, a furnace, a tannery, or an animal slaughterhouse within twenty-five meters of a well (Babylonian Talmud, Tractate Megillah 29a). From the New Testament we learn that it was recognized that polluted water led to disease: "a third of the waters became wormwood, and many men died from the water, because it was made bitter" (Revelations 8:11). Baptism in clean, running water was a means of purification. Special taxes were imposed on all inhabitants to secure a water supply for drinking and irrigation (Tractate Shekalim 1:1).

Contemporaneous with the Talmudic and early Christian era was the period of Greek ascendancy and the spread of Hellenism. Although little appears in Greek writings about water safety or purity, the Greek physician Hippocrates wrote that spring water or rainwater should be favored over stagnant water. He also wrote that "when one comes into a city to which he is a stranger . . . one should consider most attentively the waters which the inhabitants use. . . ." The Greeks' greatest contribution to public health was their strong emphasis on personal hygiene, especially regular bathing.

It was under the Romans that public health and water quality reached their acmes in the ancient world. Diligence in seeking and choosing water sources was widely practiced. By the first century C.E. the city of Rome was supplied by nine aqueducts from sources in the surrounding upland countryside, where groundwater contamination was uncommon. The aqueducts had basins at regular intervals to allow for settling of sediments, as well as sand and stones that acted as filters (much as is done today). In the city of Rome itself there were large cemented reservoirs from which water was brought by pipes to the homes of the rich or to public fountains for the masses. By the end of the third century C.E. there were more than 1,300 such public fountains. With regard to public sanitation, there was an extensive system for controlling sewage and storm water runoff in Rome, as well as a system for lowering water levels in the marshes. There were also officials responsible for caring for gutters and cleaning the streets.

Throughout military campaigns the Romans were sure to protect their sources of water. In a report written by the Roman Vegetius in the year 375 C.E., it is observed that "an army must not use bad or marshy water: for the drinking of bad water is like poison and causes plagues among those who drink it." The same author also observed that if an army were to remain camped in one place for an extended period of time during the summer or fall, it could be observed that the "water becomes corrupt, and because of the corruption, drinking is unhealthy . . . and so malignant diseases arise which cannot be changed except by frequent changes of camp." In this regard, the Romans noted that by draining marshes from which drinking water was obtained, a certain illness characterized by high fever, chills, and a large spleen could be avoided. This was an accurate description of malaria, for which marsh drainage is still practiced today for disease control purposes.

After foreign military conquests the Romans constructed aqueducts to bring water to large cities. In Carthage, for example, Hadrian built an aqueduct eighty miles long that brought millions of gallons of water daily to the city. In the time of Augustus the Romans built another large aqueduct in Nimes, France. By so doing, the Romans spread their culture of cleanliness and sanitation to the world known at that time.

Despite all of these measures, the sanitary conditions in biblical and ancient times were not adequate by today's standards. Numerous plagues are listed in the Bible and in the writings of the Greek historian Thucydides and the Roman author Pliny. Still, certain rudimentary standards were observed. Although these standards would today appear self-evident, they were largely forgotten until as recently as the end of the nineteenth century. During the intervening centuries the Western world took a step backward with regard to water quality and public sanitation.

The Middle Ages through the Eighteenth Century

With the demise of the Roman Empire and its administrative structure, the framework for public sanitation in Europe fell into disrepair. In towns and cities, the streets were littered with garbage and excrement. There were few drains or sewers, and housing was of poor quality. Filth became an ordinary part of daily life. Hygienic precautions and sanitation, both personal and public, were ignored, and disease became more common.

Christianity, which was on the rise, interpreted the increased prevalence of illness as punishment for sin or as a means of preparing the soul for redemption in the world to come. The abasement of the body was seen as a means of salvation. This included not bathing and not laundering clothing. Monks or holy men, for example, bathed only once or twice a year. The concept that disease could be prevented or contained was either not entertained or not encouraged.

It is not surprising, then, that the history of the Middle Ages is characterized by numerous plagues and forms of pestilence. During the Crusades, for example, the majority of the Crusaders died of illness along the way to the Holy Land. During the Black Death a large percentage of Europe's population perished. Diarrheal illnesses such as typhoid and cholera, which are food- and water-borne infections respectively, were also common and deadly. Although it gradually became appreciated that many epidemics were contagious and that quarantines were useful in containing outbreaks, there was little knowledge regarding the cause of these diseases.

The period of the Middle Ages was followed by the Renaissance, which lasted from approximately 1300 to 1600. Among the defining characteristics of the Renaissance was its emphasis on the dignity of man, greater personal freedom, and the opportunity to question previously held beliefs. Many of the works of classical antiquity were rediscovered. Little of this, however, led to an improved quality of life for the majority of people. Overcrowding, plagues, poor sanitation, and contamination of water persisted.

The Renaissance gave way to the seventeenth century, which was marked by great scientific advances and the development of the scientific method. Among its luminaries were Newton, Galileo, and Kepler. Antony Van Leeuwenhoek, developer of the microscope, described the microscopic world of bacteria and parasites. Again, however, little of this knowledge was translated into practical measures to improve the public's health. Even Van Leeuwenhoek's observation that rainwater was purer in content than stagnant water did not lead to any practical improvements in sanitation. It had not occurred to anyone at the time that these microbes could be the cause of illness. At the time phlogiston—a substance hypothesized to be released from a combustible material as the material burned—was still considered the cause of most illness. Epidemics were still common, sanitary conditions remained primitive, and water supplies continued to be polluted. Indeed, people at that time preferred drinking ale and wine to drinking water, owing to water's vile taste.

The first half of the eighteenth century was characterized by the Enlightenment, with an emphasis on rationalism and empiricism that allowed for further experimentation and questioning of old beliefs. The second half of the century was notable for the beginning of industrialization. Factories grew, and urbanization increased. Along with the usual diseases associated with crowding and poor sanitation, newer illnesses began to arise. Among these were lead and mercury poisoning, as well as respiratory illnesses due to the smoke from factories burning coal. Toxic wastes were dumped into rivers situated next to factories, further polluting rivers and estuaries. Such wastes, of course, were present in addition to the human wastes that were regularly being discharged into these bodies of water all along.

Although medical advances made during this time, such as vaccination (introduced by Edward Jenner), were important in containing the spread of disease, there was still no overall approach to public health or sanitation. It was only in the nineteenth century that these concepts were developed and that public health—including water quality—became important.

The Nineteenth Century

The concepts of public health and sanitation had their origins in the nineteenth century. They arose from two different sources. The first was the growing realization, in the first half of the century, that government had a moral and ethical responsibility to protect the welfare of its citizens. Much of this realization developed as a result of the Enlightenment and the democratic movements in the

United States, England, and France during the previous century. For example, Jeremy Bentham, an English philosopher, advocated numerous governmental reforms to address social and health problems. The French philosopher Octave Mirabeau proclaimed that the health of the citizenry was the responsibility of the state. As a result of these new concepts, a greater number of legislative deliberations and official inquiries regarding sanitation were conducted. During these deliberations emphasis was placed on ensuring an adequate supply of wholesome water to the poor, since it was increasingly recognized that epidemics and poverty were closely linked. In London settling reservoirs and filters were introduced to remove gross particulate matter from the water derived from the Thames River.

The other source from which the concepts of public health and sanitation arose in this period was the development of organic chemistry and microbiology. Public health and sanitation were now supported by a scientific method that was quantitative and reproducible and presented a cohesive framework for inquiry. Scientists such as Louis Pasteur and Robert Koch showed that the organic compounds in water were microorganisms called bacteria, that these bacteria were derived from human or animal waste, that these microorganisms caused disease even in low concentrations, that water was a temporary habitat for many microorganisms and served as a medium of transfer from an infected person to another who was healthy, and that removal of microorganisms from water supplies could lead to the control and prevention of epidemics. These discoveries allowed for a major leap forward for the health of the community.

Armed with this knowledge, cities undertook sanitary measures. The first milestone in this regard followed the demonstration in 1854 by the English physician John Snow that a cholera epidemic in London could be traced back to a single source of water. His evidence was compelling, and when he removed the handle of the water pump on Broad Street, thereby shutting down the source of water infection to the public, the epidemic abated and was stopped. Soon public water commissions were set up and uniform standards for water purification were developed. Public cesspools and latrines, as well as the dumping of untreated human and animal waste and of animal entrails from slaughterhouses into rivers from which the community derived its water, were stopped. For the first time in nearly two thousand years, Europe enjoyed water of a decent quality. Consequently, by the end of the nineteenth century many water-borne infectious outbreaks began to decline in prevalence.

2

The Modern Era
of Drinking Water
Regulation

The modern era of drinking water protection and regulation is most strongly reflected in developments in the United States. It is therefore appropriate to relate the history of drinking water regulation in the twentieth century by describing the events and legislation that have affected drinking water in this country.

The First Half of the Twentieth Century

In the United States in the nineteenth century, as in Europe, there were numerous outbreaks of diarrheal disease due in part to contaminated water. In New Orleans there was a large cholera epidemic in 1833. In Chicago and Philadelphia, typhoid was endemic. Indeed, diarrheal diseases were the third leading cause of death in the United States at the turn of the twentieth century.

The discovery that these diseases were infectious and transmissible through water served as an impetus to the development of methods of water purification. In Lawrence, Massachusetts, the first U.S. water filtration system was established in 1887 to sift the waters coming from the Merrimack River. Sand was the filtering medium. Within a short time the incidence of typhoid fever dropped precipitously. This success led to the introduction of several other types of mechanical and sand filters for municipal water supplies throughout the United States. By 1907 thirty-three cities in the United States had installed mechanical filters and thirteen others were using sand filters. Concurrent with these developments, in 1899 the Congress enacted the Rivers and Harbors Act, prohibiting the discharge of refuse into navigable waters.

Several years after the introduction of filtration, another method of water purification was introduced—chlorination. Chlorine had recently been found to reduce the number of microorganisms in water by diffusing through the cell walls of the infecting agents and poisoning them. Chlorination was first used in Jersey City, New Jersey in 1909. The city's reservoir would at times become contaminated from river sewage. Rather than introduce new filtration systems, the water company decided to chlorinate the water supply. It soon became clear that the bacterial counts in the water had dropped to negligible levels and that chlorination was an effective method of ensuring water purity. Soon chlorination became an additional and commonly used method of water purification. In the United States today, chlorine is used in 75 percent of the larger municipal systems and 95 percent of the smaller systems.

In 1948 the Federal Pollution Control Act was passed into law. With subsequent amendments, it became the Clean Water Act. It enabled the government to more forcefully implement the prohibition of pollutant discharge into navigable waterways. In particular, the act regulated producers of pollution, such as factories and sewage treatment plants, and required them to pre-treat their effluents to minimize water pollution at its source.

As a result of these measures, as well as the establishment by the government of bacteriologic standards for water quality, the common infectious causes of water-borne illness became well controlled. Major outbreaks of typhoid, cholera, and other diarrheal diseases became rare occurrences or were quickly contained. It could be stated that by the middle of the twentieth century the U.S. public was enjoying a high quality of drinking water as never before.

The Second Half of the Twentieth Century

The second half of the twentieth century saw two new developments with regard to drinking water. The first was the use of drinking water as a means to promote public health. This was done through the introduction of fluoride into public drinking water supplies for protection against dental caries (cavities). This was the first time that drinking water was used in a proactive manner.

The other major development of the latter half of this century was the realization that drinking water sources were being contaminated by chemicals and that this contamination was injurious to health. The industrialization of the United States following World War II had led to a marked increase in the use of synthetic chemicals. The problems associated with disposal of these chemicals soon became noticeable when the levels of chemical pollutants in water sources rose. The threat posed by these chemicals was first noticed in the fish and mammals that lived in or near these water sources. Finding dead fish with disfiguring tumors in

inland waters had a profound effect on the public conscious-
ness and on legislative bodies. Dying sea mammals near
ocean coasts also caught the public's attention. The growing
scientific awareness of the adverse effects of toxic substances, the fear of cancer, the enhanced ability to measure
chemical levels through improved technologies, and the rise
of the environmental movement made it clear that the regulations of the 1940s were no longer adequate to protect the
drinking water supply in the United States.

In response to these challenges, Congress enacted the
Safe Drinking Water Act (SDWA) of 1974. This act was the
most comprehensive in history to protect water sources and
drinking water. It set up government oversight, through the
Environmental Protection Agency (EPA), of surface and
groundwater sources, programs for the development of standards and regulations, and funding for state water systems.
Ongoing monitoring to ensure compliance was made an
integral part of these programs. National standards for protection from harmful contaminants were authorized so as to
"protect to the extent feasible, using technology, treatment
techniques, and other means, which the Administrator
determines are generally available (taking costs into consideration)."

To attain these goals, the EPA set up two types of regulations: (1) mandatory, enforceable maximum contaminant
levels (MCLs), to be set as close to the recommended health-
based goals "as is feasible"; these levels would pose no significant health risk if drunk over a lifetime and (2) non-
mandatory, health-based maximum contaminant level goals
(MCLGs) established for each toxic substance that "may have
an adverse effect on the health of persons . . . allowing for an
adequate margin of safety."

Several classes of chemicals and contaminants were to
be regulated:

Microbiological Contaminants. These include bacteria, viruses, and parasites. They usually originate at drinking water sources. A certain minimum level of purity is required to prevent epidemic outbreaks of diseases.

Metals and Inorganic Chemicals. These include lead, copper, mercury, and aluminum, as well as non–carbon-containing chemicals such as sulfates, nitrates, nitrites, and asbestos. Some of these enter drinking water at its source from the local geology from which the water originates; others enter water via the conduits (pipes) through which water is distributed. These substances can affect the nervous system, lead to cancer, or cause blood disturbances.

Volatile Organic Chemicals (voc). These are carbon-containing chemicals that vaporize easily from water to air. They derive chiefly from solvents, insecticides, industrial wastes, and underground chemical storage tanks and reach water sources through intentional and unintentional pollution. These chemicals include benzene, carbon tetrachloride, and heptachlor. They have been associated with cancer and impaired fertility.

Organic Compounds. This category includes several large classes of carbon-containing chemicals. Among them are herbicides and pesticides that are directly and intentionally applied to soil that may be near water sources, chemicals intentionally or unintentionally spilled into the environment by industry and individuals, and secondary organic contaminants that result from organic compounds that combine with chlorine and by-products of chlorine disinfection. These substances have been associated with impaired fertility and cancer.

Radionuclides. These include naturally occurring isotopes of uranium, radon, and radium that enter the water supply through the local geology from which the water originates. They can be associated with cancer.

We describe each of these categories of contaminants and their effects on health in greater detail in the coming chapters. In Appendix 1 is a complete list of these contaminants.

The EPA derived these maximal allowable levels based on two standards. The first was an immediate adverse outcome standard. For example, if water was polluted with bacteria, diarrhea developed. The need for regulation to prevent such short-term effects was obvious. The other standard was to prevent longer-term outcomes of pollution. In 1974 it was felt that the worst possible outcome of pollution was the development of cancer, and for that reason guidelines were established with cancer prevention in mind. Through various surveys the EPA established that society as a whole would be willing to pay two to $10 million to implement regulations that controlled levels of potential cancer-causing agents (called carcinogens) in water that would, in theory, "prevent" one person in a population of one million people from developing cancer as a result of exposure to that substance in drinking water over a period of seventy years. This expensive, conservative approach was adopted since exposure to potential carcinogens through drinking water is *involuntary*, and an individual exercises no control as to whether he or she is to be exposed to a potential carcinogen in drinking water.

The problem with this approach is twofold. First, it is very expensive. The cost may not be so strongly felt by large municipal water suppliers. On the other hand, the cost of detection, treatment, and monitoring is relatively much more expensive for small water suppliers. As of 1990 there were more than 51,000 small water suppliers in this country serving populations of fewer than 3,000 people.

The second problem is that the emergence of newer illnesses and medical concerns has made the older standards less relevant. As we detail in the coming chapters, problems as diverse as human infertility, heart disease, and deteriorating memory (dementia) may be related to the contaminant content of drinking water. These effects are subtle and take longer to come to the fore than does the development of cancer. For that reason, newer regulations for water quality are and will be necessary.

What is clear is that the regulation of water purity is an ongoing process, and its effects on health will require constant vigilance. The fact that the SDWA exists and provides a mechanism by which to continually monitor drinking water purity and to amend existing legislation assures us that the water we drink will remain of high quality.

3

Drinking Water Sources, Treatment, Safety, and Conservation

The world's supply of freshwater, from which drinking water is derived, is constant. The world's population and industrial commerce, on the other hand, are constantly increasing, and the demand for freshwater is growing. It follows that drinking water, which already is a somewhat scarce commodity, will with time become even scarcer.

Today the United States and the rest of the industrialized world enjoy large supplies of high-quality drinking water. Not so the rest of the world. According to the World Health Organization (WHO) and the United Nations Children's Fund (UNICEF), nearly half the world's population of approximately six billion people does not have access to proper hygienic sanitation systems. More than 1 billion people have no access to safe drinking water. Almost all of these problems are seen in the developing world. The resulting diarrhea and illness lead to poor health and impaired economic development. In 1996 alone, more than 2.5 million deaths occurred from diarrhea, mostly in children. Infectious diseases still account for the majority of deaths (approximately 45 percent) in the developing world, part of which is due to contaminated drinking water.

The fact that many of the water quality problems present in the developing world are no longer present in this country does not mean that safe drinking water should be taken for granted. The supplies of drinking water are not as bountiful as they once were. In southern California and the southwest, control of water rights and water catchment areas has become a contentious issue, with many economic, agricultural, and political interests at stake. New York City has recently taken steps to protect its upstream water sources. Other cities such as Atlanta, New Orleans, and Cincinnati that derive their water from river sources have had to cope with ever-increasing levels of pollution. In cities such as San Juan, Puerto Rico, where municipal water supplies are derived from reservoirs, the demand for water has occasionally outstripped the supply due to population growth. Twice in three years, San Juan has required water rationing. All of us are aware of local government ordinances not to water lawns or wash cars in times of drought.

In this chapter we give an overview of drinking water sources. We also discuss how "raw" water is processed and treated so as to become fit for drinking, and we describe the methods used to guarantee water safety. Last we will describe the overuse of water and what is being done about it.

Sources of Water

In this country drinking water is almost exclusively derived from surface waters and groundwaters. Surface waters include lakes, reservoirs, and rivers. Groundwaters are aquifers into which wells have been drilled. Aquifers are underground geological formations that contain water. The source of one's water will vary depending on where one lives. For example, cities built on rivers derive their waters from the rivers. Those built on lakes use the lake waters. Other cities that are not on rivers or lakes may rely on large underground aquifers or transport water from distant watersheds (areas in which water naturally collects).

For the most part, large municipal water suppliers tend to use surface waters, whereas smaller suppliers use groundwaters. Private groups or individuals usually use groundwaters as sources of drinking water. Altogether, slightly more than half the population uses groundwaters as their sources of drinking water. If one is unsure of the source from which one's drinking water is derived, one can contact the local water utility company or public works department. Table 1 is a list of the water sources of the largest municipalities in the United States.

Other potential sources of water are the oceans and rain. Removing salt from ocean water (desalination) is an expensive process that is not used in this country. Collection of rainwater in cisterns is occasionally done in remote parts of the United States. Rainwater is an important source of water in places such as the U.S. Virgin Islands, Bermuda, and the peninsula of Gibraltar in the Mediterranean.

Water Treatment by Public Utilities and Water Distribution

Water utility companies use a variety of processes to remove contaminants from "raw" water to make it drinkable. This is the process of water treatment. These processes include flocculation, sedimentation, filtration, adsorption, and disinfection. Rather than describe each of these processes

Table 1. *Water Sources of the Largest Municipalities in the United States*

Municipality	Source
Atlanta	Chattahoochee River
Boston	Quabbin Reservoir
Chicago	Lake Michigan
Columbus	Hoover Well and Griggs Reservoir
Dallas	Lakes Ray Hubbard and Tawakoni
Denver	South Platte River
Detroit	Detroit River
Houston	Trinity River and Lake Houston
Los Angeles	Stone Canyon, Encino, and other reservoirs
Miami	Two large aquifers, part of the Floridian Aquifer
New Orleans	Mississippi River
New York City	Schoharie, Ashokan, and Pepacton Reservoirs
Philadelphia	Delaware and Schuylkill Rivers
Phoenix	Salt and Verde Rivers
Portland	Bull Run Lake
St. Louis	Missouri River
Salt Lake City	Little Cottonwood Creek and Deer Creek Reservoir
San Diego	Colorado River
San Francisco	Sunset and University Reservoirs
Seattle	Told and Cedar Rivers
Washington, D.C.	Potomac River

separately, we will describe how the water of one community—Atlanta, Georgia (our hometown)—is treated. The reader can then extrapolate the details to his or her own community.

The Chattahoochee River runs through the Atlanta metropolitan area. At its origins in the southern Appalachian Mountains it is pristine. As it flows nearer and nearer Atlanta, however, it becomes progressively polluted. When one cultures the water from samples taken at different points along the river, one finds that the number of bacteria per milliliter progressively rises. This is called the coliform count, and it is a marker of organic (carbon-containing) and fecal material in water. Atlanta is not unique in this respect. In fact, there is no expectation that a municipal water source will be pure. This is the purpose of a water treatment facility.

As the first step in water processing in Atlanta, water is pumped out of the Chattahoochee River at a rate of thirty to thirty-five million gallons per day. To rid the water of gross debris it is first passed through a large metal screen. Next the chemical potassium permanganate is added to initiate the sterilization process. After this, four other chemicals are added to promote further sterilization. First, chlorine in gas form is bubbled into the water. This is the primary means used to kill bacteria. Second, orthophosphate is added to protect the plant's pipes from corrosion from the many organic and acidic substances in the untreated water. Third, alum (aluminum sulfate) is mixed into the water to act as a coagulant. It gathers particles present in the water into large clumps that later settle out. Finally, carbon is used to absorb tastes and odors. Many of the organic substances found in untreated water do have an odor.

The water is then allowed to dissolve these ingredients for fifteen to twenty minutes. It is churned in a chamber with huge rotating paddles called the flash mix. From there it flows into large concrete vats known as flocculation basins. Each of these is larger than a community-sized swimming pool, and they are arranged in series. In each basin the water moves progressively more slowly. It takes from twenty to sixty minutes to proceed through all the basins. Peering into these basins, one sees tiny particles congealing near the surface (due to the alum), growing progressively heavier, and then falling toward the bottom as their density increases. This process is called flocculation. The water is then skimmed through a weir (named after its inventor, John Weir), a saw-toothed device through which the water passes, leaving behind the flocculation particles (which are heavier and have fallen deeper than where the weir is placed). The water that remains in the final flocculation basin is detoured to a sedimentation tank, where it sits for six to seven hours.

The "muck" that eventually settles at the bottom of the sedimentation tank is collected and carted off to a solid waste facility; the remaining water is reprocessed. The processed water, minus its sediment, proceeds from the weir to filtration tanks. Atlanta has thirteen of these tanks, all of which are located indoors. Water flows into these room-sized tank containers from the bottom up, percolating through progressively finer filtering materials: gravel, sand, and then charcoal (anthracite). By this point in the process, the water is clean. Post-treatment chemicals are then added, specifically fluoride and chlorine gases. From here the water travels to an underground clear well that holds 14 million gallons of water.

Adjoining the Atlanta treatment facility is a laboratory. In it drinking water is checked to ensure that proper sterility and chemical balance have been achieved. The monitoring equipment that is used is called a Milton Roy pilot filter, and it duplicates the entire process of water purification in miniature. The water's cloudiness (called turbidity) is the primary measuring characteristic used to judge the effectiveness of the water treatment process. On the day of our visit, the Milton Roy device measured the water's turbidity at entry at 44.3 NTU or turbidity units. Upon completion of the treatment process the water's turbidity was only 0.4 NTU! The Milton Roy equipment directs the plant's engineers to alter specific plant operations to achieve the water purity that is sought. Also in the laboratory, the chlorine concentration of the water is continuously monitored and its coliform counts are periodically monitored.

From the plant the water is transported under pressure through a distribution network of buried pipes. To these main pipes smaller service pipes are attached by which water is brought to individual homes or offices. In many communities water is first pumped to storage tanks that are located at elevations higher than the homes they service. Gravity is then used to provide the pressure that allows the water to flow when a tap is turned to open a pipe.

The Safety of Community Drinking Water

There are nearly 55,000 community water systems in this country, supplying water to meet the drinking needs of more than 90 percent of the U.S. population. The Environmental Protection Agency (EPA) defines a community water source as one that consistently provides water to at least twenty-five people for at least sixty days during the course of a year.

Information on water quality in one's area is available from the local public health department or water supplier. State agencies also have departments responsible for drinking water quality. Finally, the EPA in Washington and its ten regional offices may also provide information (available on the world wide web at www.epa.gov).

In 1996 approximately 7 percent of community water systems were reported to have exceeded one or more maximum contaminant levels for the eighty substances monitored by the EPA (see Appendix 1 for details). Some were minor, whereas others were more serious. A recent survey reported in *USA Today* (October 21, 1998) suggests that the number of water systems that are not in compliance with federal regulations may be even higher. In 1996 Congress passed several amendments to the Safe Drinking Water Act (SDWA) to ensure water safety. These amendments specified, among other things, that users of public water systems are to be informed about the quality of their drinking water and what is being done to protect it. Suppliers of drinking water are now required to inform the public within twenty-four hours if drinking water has become contaminated with anything that might cause immediate illness. Potential adverse effects on human health, the need to use alternative water supplies until the problem is solved, and what steps are being taken to correct the problem must be announced through the media. Those violations that do not cause immediate harm can be reported in water bills the month after the infraction or in letters delivered through the mail. Beginning in 1998, each state began to compile information on individual water systems so that comparisons between systems can be made. This information, in turn, is forwarded to the EPA, which will compile comprehensive annual reports on the quality of the nation's drinking water. Beginning in 1999, individual water systems are required to prepare and distribute to their customers annual reports that detail the source from which the supplier derives its water, results of monitoring during the course of the previous year, and information regarding potential health hazards associated with any detected violation of the EPA standards.

To ensure proper water testing, the EPA has established a schedule for pollutant testing. The pollutants and the frequency of testing include:

- Bacteria (e.g., coliforms)—monthly or quarterly, depending on system size.
- Viruses and parasites (e.g., *Giardia*)—monthly; continuously for turbidity.
- Nitrates—yearly.
- Volatile organic chemicals (e.g., benzene)—annually.
- Synthetic organic chemicals (e.g., pesticides)—twice every three years for large systems, once every three years for small systems.
- Metals (e.g., mercury)—once every three years for groundwater; annually for surface water.
- Lead and copper—annually.
- Radionuclides (e.g., radon)—once every four years.

If in the course of testing it is found that contaminants exceed the maximal levels, consumers must be notified and immediate reparative measures must be taken. Retesting is mandated until it can be shown that the problem has been fixed.

The 1996 amendments to the SDWA provide several additional assurances of the protection of water quality. States are now required to develop programs to identify potential problems that may affect water sources before they happen. In this way it is hoped that problems will be prevented. Each state is also required to train and certify the individuals who operate drinking water utilities. The 1996 law also provides nearly $10 billion for improving the drinking water infrastructure over the next six years. These provisions will be administered through several government agencies, including the U.S. Department of Agriculture Rural Utility Service, the Department of Housing and Urban Development, and the Economic Development Administration.

Finally, a Partnership for Safe Water was formed in the mid-1990s. It is a voluntary cooperative effort between the EPA, the American Water Works Association, and nearly 250 water treatment plants. Its goal is to enhance compliance with water standards throughout the water industry and to analyze and research questions that remain unresolved regarding water quality. Such a program ensures ongoing quality control for the water industry.

In case there is a toxic spill or a treatment problem is detected—or one is simply unsure of the quality of his or her water supply—there are several things that he or she can do. The first is to boil water for drinking. This is effective in ridding the water of microorganisms, though it may not improve its taste. One can also install a filtering system for added safety or to remove unwanted tastes and odors (see Chapter 10). Finally, one can buy bottled water (see Chapter

9). Both filtering systems and bottled water are under government regulation, so high-quality water is ensured.

Safety of Private Drinking Water Supplies

Nearly 23 million Americans derive their water from private drinking water supplies. Most of these people live in rural areas, and the majority use water drawn from underground sources, such as wells. Newer wells are drilled down to the level of an aquifer. Electric pumps and gauges provide the pressure to push the water up into individual homes. In the home, the water is filtered to remove particulate and organic matter, acids, and minerals. Older wells are usually dug and do not extend as far below the surface as newer wells. The water is generally accessed through an above-ground pump. Streams and rain water stored in cisterns are other sources of water.

These water sources do not fall under the aegis of government regulation. Therefore the onus for testing falls directly on the user. Annual testing for coliform bacteria is recommended. Because many private wells are near farms where nitrate-rich fertilizers are intentionally applied to the soil, testing for nitrates is also strongly recommended. In areas with high levels of background radioactivity in the local geology, radionuclides, such as radon and uranium, should be tested for as well. Water from metal cisterns should be tested for metal concentrations.

Many commercial laboratories are available for testing. Lists are available from state departments of health or regional EPA offices. Some local health departments do water testing as well. Basic testing costs from $10 to $20. The laboratory generally supplies the sample bottles, with special instructions for handling. The samples are mailed in, and within several days a report is generated. If contaminant levels are above a certain critical level, the consumer is advised to retest the water and to contact his or her local health department for information on and assistance in proceeding.

Perhaps the easiest and most effective means of safeguarding one's private water supply is to protect it. That means to keep hazardous chemicals and human and animal wastes as far from the water source as possible. Assistance in this regard is available from the U.S. EPA Safe Drinking Water Hotline (telephone: 800 426-4791). For farmers there is an organization called the Farm*A*Syst/Home*A*Syst Program (address: B142 Steenbock Library, University of Wisconsin, Madison, Wisconsin 53706; telephone: 608 262-0024; e-mail: farmasyst@macc.wisc.edu; web site: www.wisc.edu/farmasyst).

Use and Overuse of Drinking Water and Efforts at Conservation

The greatest benefits of municipal water are its easy availability, its low cost, and its long track record of safety. The addition of fluoride to most community sources is an added benefit. The provision of water generally is a government service that residents pay for over and above taxes, but nonetheless represents a considerable bargain. On average, a quart of community water costs approximately a tenth of a penny. Perhaps because of this, Americans use more drinking water per person per day than the rest of the world. A typical family of four uses approximately 350 to 400 gallons per day. In contrast, in Belgium or Canada a family of four may use only 40 percent or 60 percent as much water, respectively. A typical American household that derives its water from a private well is likely to use only 200 gallons per day.

Where does all this water go? Only a small portion goes toward drinking. The vast majority is used for bathing, flushing toilets, cooking, washing clothes and dishes, and preparing food. Watering lawns and cleaning cars are other uses. Additional water is used in the community for park maintenance, street cleaning, and firefighting. Industry also makes demands on water use, especially in certain defined areas. For example, it takes 7 gallons of treated water to process one bushel of corn to sweeten 400 bottles of soft drinks, whereas making a ton of Belgian chocolate requires only 4 gallons of purified water. On the other hand, in electronic manufacturing tens of gallons of purified water may be required for one rinse cycle for one microchip!

As we approach the twenty-first century and the challenge of accommodating a projected 10 billion people by midcentury, the issue of water conservation will take on greater urgency. As we mentioned earlier, the total amount of freshwater from which drinking water can be derived is fixed. Only 3 percent of the earth's water is fresh, and of that less than 1 percent is accessible. It is not surprising, then, that freshwater and drinking water are, and will continue to be, market commodities that one day will be much more expensive than they are today.

In order to conserve water, several measures have recently been adopted or proposed:

- Using water-efficient faucets, toilets, and showerheads.
- Enforcing building codes requiring the use of water-efficient appliances and fixtures.
- Surveying for leaks and testing meters in buildings.
- Improving lawn-watering techniques.
- Recycling industrial water.

Many government and industry initiatives have been undertaken lately to enhance water conservation. The 1996 amendments to the SDWA specified that the EPA was to draft guidelines within two years (by 1998) "for water conservation plans for public water systems." These guidelines are actively being developed at present, and public meetings are being held to formulate rules. Minutes of the meetings are available at the EPA web site (www.epa.gov).

An outgrowth of this conservation plan is WAVE. Water Alliances for Voluntary Efficiency is a nonregulatory water efficiency partnership created by the EPA to encourage commercial businesses and institutions to reduce their water consumption while increasing their efficiency and profitability. Among the charter members of WAVE are members of the hotel and lodging industry, including names such as Sheraton, Hyatt, and Westin. The members of this organization believe that water consumption can be reduced by 30 percent through proper and intelligent use of resources. The typical payback period for the initial capital outlay required for the necessary equipment is three years.

There is reason to be optimistic that these efforts will bear fruit. Some results are already evident. In December 1998 the U.S. Geological Survey reported that Americans' use of water had declined by nearly 10 percent from 1980 to 1995. The population increased by 16 percent during the same time period. Thus measured on a per person basis, water consumption decreased by 20 percent in this time period. This is particularly striking because water consumption increases had greatly exceeded population increases in all previous reports.

According to the 1998 report, the greatest relative reductions in use were in agriculture and industry. More efficient irrigation systems, the ability to recycle water inexpensively, and efficient new manufacturing technologies all helped to make this reduction possible. The greatest increase in water use occurred in the livestock industry, reflecting its growth over the past decades. On the other hand, domestic use of water remained stable at 100 gallons per person per day. The Northeast and Midwest showed the greatest declines in water use, followed by the Plains and the West Coast.

The most important way to ensure future supplies of drinking water, however, is to prevent the contamination of freshwater sources. Since 1972 and the enactment of the Clean Water Act, the progress in this regard has been impressive. The number of Americans served by sewage treatment facilities has doubled. The number of lakes and rivers that are safe for fishing and swimming has doubled (from approximately 30 percent to 60 percent). National clean water standards have prevented the disposal of billions of tons of industrial pollution into freshwater sources.

Despite these giant strides, much work remains to be done. States report that up to 40 percent of freshwaters surveyed are still too polluted for fishing and swimming. Controls for pollution from factories and sewage treatment plants have not yet matched the controls in place for mining, forestry, and agriculture. To address these issues and to reinvigorate the water conservation movement, a new Clean Water Action Plan has been proposed. Although a detailed discussion of its contents is beyond the scope of this book (see the EPA web site, www.epa.gov/cleanwater/action), its guiding principles are to promote cooperation between various federal, state, and local governments and private industry and to avoid duplication of action; to enhance community participation; and to emphasize innovative, market-based approaches to pollution control. Its goals are to enact and enforce clean water standards; to enhance watershed management; to restore watersheds that do not meet clean water goals; to build bridges between water quality and natural resources programs; to respond to growth pressures on sensitive coastal waters; to prevent polluted runoff; to protect from pollution the 800 million acres under federal government control, many of which are headwaters for rivers and lakes; and to improve information about water and enhance means for citizens to exercise their right to know.

Ultimately, for this plan to work and to safeguard the availability of clean water in the future, the broad public and industry must both be involved, and everyone's cooperation will be needed.

Part Two

Drinking

Water and

Disease

4

Drinking Water
and Infectious
Diseases

Disease-producing bacteria were the first contaminants of drinking water recognized to pose a threat to health. As a result of this recognition, wide-ranging sanitation and purification measures were undertaken. This has led to a remarkable reduction in infectious diseases due to contaminated drinking water in the Western industrialized world.

Despite this impressive accomplishment, water-borne infectious diseases remain an issue of concern. From 1971 to 1988 there were 564 water-borne infectious outbreaks in the United States involving nearly 140,000 people. In 1993 and 1994 there were five such outbreaks in four states. More than 400,000 people became ill, and nearly 50 died. The incidence of diarrhea due to water-borne infection remains substantial. Each year in the United States more than 200,000 children under age five are hospitalized for diarrhea, part of which stems from ingestion of contaminated water.

In addition to the ongoing concern about well-known water-borne infectious agents, there is now concern about several newly recognized water-borne infectious agents. The control and prevention of these agents has become a large challenge to public health.

This chapter is divided into three sections. The first deals with bacteria; the second with parasites; and the third with viruses. In each section we touch upon the public health impact associated with these infections. Water-borne worm infections are not discussed owing to their rarity in this country, though they are a large problem worldwide.

One note regarding methods. The statistics reported here are mostly derived from the Centers for Disease Control and Prevention (CDC), the world's premier public health agency. The data used are inexact owing to many inadequacies of reporting. Reporting from states and localities, upon which national data depend, is often incomplete. Then, too, many outbreaks are not investigated. In many instances no causative agent is found. Therefore, national figures are approximations only. However, they are the best estimates that are available and probably paint a relatively accurate picture of trends.

Bacterial Infections

Bacteria are microorganisms that are of the kingdom *Prokaryotae*. They are like microscopic single-celled plants. They are either round, rodlike, or spiral, and are characterized by having a cell wall or outer membrane. They often aggregate in colonies. They are ubiquitous and live in soil, water, and organic (carbon-containing) matter. They live inside the intestines of animals, including man, where they help with the process of digestion. When certain bacteria appear in places where they do not normally reside, they can cause illness. If these bacteria are ingested, as in drinking water, they may cause illness, usually diarrhea and abdominal cramps.

Cholera

The prototypic water-borne bacterial disease is cholera. It is an acute intestinal infection caused by the bacterium *Vibrio cholera*. It is spread by eating food or drinking water contaminated by the fecal waste of an infected person. This occurs most often in underdeveloped countries lacking pure water supplies and proper sewage disposal. In this country it is most likely to happen when raw shellfish from polluted coastal waters are eaten. Most persons infected with this bacterium do not become ill. When illness does occur, it is generally mild. In 5 percent of cases, however, severe diarrhea and dehydration occur, and if not properly treated, it can result in death.

Cholera was known in the classical world. It was common in the Western world in the last century. Tens of thousands died of cholera epidemics in Germany, England, and the United States until the middle of the nineteenth century. Not until 1854, when John Snow removed the handle of the Broad Street pump in London, England, making contaminated water unavailable for distribution (see Chapter 1), was the means for the spread of cholera understood. Today cholera is a rare illness in the developed world due to modern sewage and water treatment facilities. Nonetheless, it is still a major problem in the developing world. Owing to modern travel and commerce, the potential for spread to this country remains present.

In 1961 a worldwide cholera epidemic (called a pandemic) began when the *Vibrio* bacterium appeared in Indonesia. The disease soon spread to eastern Asia, and then to Bangladesh in 1963, to India in 1964, to the U.S.S.R., Iran, and Iraq in 1965 and 1966. In 1970 it reached West Africa, which had not experienced cholera for more than 100 years. In 1991 cholera reached tropical Latin America. Within the year it spread to eleven countries, and thereafter to the rest of the continent. From January 1991 to September 1994 more than one million cases and more than 9,600 deaths (a fatality rate of 0.9 percent) were reported in the western hemisphere. This new phase of the epidemic produced as many as 4,500 new cases per day in Peru. During the same period more than 100,000 cases were reported in southern Asia. In the United States during 1993 and 1994 22 and 44 cases, respectively, of cholera were reported to the Centers for Disease Control and Prevention. Of these, 65 were associated with foreign travel.

Predicting how long the epidemic in Latin America will last is impossible. The epidemic in Africa and Asia shows no sign of abatement. Major improvements in sewage disposal and water treatment will be needed until the spread of cholera can be overcome. In this regard, we should note that improper sanitation and sewage are also problems for the United States, albeit indirectly. Many of the Mexican border towns that share a common border with Texas and Arizona are too poor to process their waste waters. Raw sewage is dumped into the Rio Grande, which threatens the water supplies of contiguous states. As a result the Arizona Department of Health Services needed to warn the public in August 1997 to thoroughly cook shellfish after laboratory testing found cholera bacteria in a sample of raw shrimp. One case of cholera occurred that year in the Phoenix area. Many Texas cities are now regularly testing drinking water sources for the cholera bacterium. Chlorination of drinking water is an effective means of eliminating contamination, but not all community water sources are chlorinated.

Persons traveling to cholera-affected areas should not eat food that has not been properly cooked (especially fish and shellfish) and should drink only water that has been boiled or treated with chlorine, iodine, or other effective germicides. Although there are vaccines against cholera, their efficacy is not high and duration of protection is short. Therefore, preventing ingestion is the key. Anyone with watery diarrhea who has returned from a cholera-affected area should seek medical attention and cases should be reported to local and state health departments.

E. coli

Bacteria that normally live in the intestinal tracts of animals and humans are called coliform bacteria. The best known of these is *Escherichia coli,* or *E. coli* for short. In the intestine, *E. coli* lives harmoniously with its host, helping with the breakdown and fermentation of food. When drinking water becomes contaminated with fecal material, *E. coli* is detected and the total coliform bacteria count of the water rises.

In 1990 the EPA promulgated an enhanced version of its long-standing total coliform rule. This legislation mandates periodic inspections of water delivery systems for bacterial infection, and for *E. coli* in particular. If water samples are found to contain bacteria, additional testing and action are to be undertaken until the problem is corrected. In such a way it is hoped that the United States can minimize disease caused by water-borne microbial transmission.

There are hundreds of strains of *E. coli,* most of which are harmless. However, when one of these strains undergoes change or mutation, it can become a source of great medical and public health concern. Such apparently is what has happened to a strain of *E. coli* called O157:H7. It gives off a potent chemical called an exotoxin that can cause severe bloody diarrhea, kidney failure, and even death.

Isolates of O157:H7 were first implicated in illnesses in 1982. Since that time there have been numerous outbreaks. Yet it was only in 1993 that this strain of *E. coli* received public attention. In that year more than 700 people in four different states became infected from it, of whom 51 developed temporary kidney failure and a low blood count (called the hemolytic uremic syndrome), and four died. The cause of this outbreak was traced back to the consumption of undercooked beef at a regional fast food restaurant. Apparently during the slaughtering process bacteria from the cow's intestine were mixed into the meat and persisted after the meat was ground up. In 1996, the first year of national reporting on this bacterium to the Centers for Disease Control and Prevention, there were 2,741 cases of reported infection, of which 102 cases were associated with kidney failure and a low blood count. It is now believed that there may actually be more than 20,000 cases per year of O157:H7 infection in the United States, resulting in approximately 250 deaths per year.

More recently it has become clear that *E. coli* can be transmitted through means other than meat ingestion. One is through ingestion of unpasteurized milk. *E. coli* can live on the udders of cows, and if the milk of these cows is not treated, one can become infected by drinking it. Another source of infection is unpasteurized apple cider, which caused an outbreak in the western United States in 1996. Drinking water has been implicated as another source of *E. coli* infection. The first and largest water-borne infection associated with this strain of bacteria was in a Missouri town in 1989. In that year water pipes froze and leaked. Sewage water was sucked into the drinking water. As a result, 240 people became infected with *E. coli*. Thirty-two were hospitalized, and 4 died. In 1991, 59 people became infected in Portland, Oregon, after swimming in a lake.

Ingestion most probably occurred when the bathers swallowed fecally contaminated water. The fact that water from so large a body of water with a very dilute concentration of the bacteria could cause disease suggests that the number of bacteria necessary for infection is small. In the summer of 1998 an outbreak occurred among 26 small children who visited a water park in an Atlanta, Georgia, suburb.

Although these outbreaks of water-borne *E. coli* infections are uncommon, they point out that *E. coli* can pose a risk. For that reason one should drink only treated water and avoid water with any possible fecal contamination. When in doubt, boiling is recommended. One should also avoid bathing in any unchlorinated freshwater that is potentially contaminated, such as water in which small children wearing diapers have played. Swimming pools are safe if the chlorine level is 1.5 to 3.0 parts per million and the pH is 7.2 to 7.6.

Campylobacter, Salmonella, and *Shigella*

We should mention several other bacteria that can cause diarrhea through water-borne contamination. Bacteria by the name of *Campylobacter* were recognized as a cause of diarrhea in 1972. Since that time an increasing number of cases have been recognized. In one study of hospital patients with diarrhea, *Campylobacter* accounted for 4.4 percent of the cases. Between 1978 and 1986, 57 outbreaks of *Campylobacter*-associated diarrhea were reported to the CDC. Of these, 11 were water-borne outbreaks. All were associated with untreated surface water or water that was inadequately chlorinated. The remainder of the infections were food-borne. In 1993 and 1994, *Campylobacter* was implicated in three of thirty water-borne disease outbreaks.

Salmonella, the bacteria that cause the diarrheal disease called typhoid, is generally transmitted through fecal contamination of food. Most commonly implicated are meat, poultry, dairy products, and eggs. Rarely, water is the means of contamination. In 1993 more than 600 people became sick from *Salmonella* in a town in Missouri when the water supply became contaminated, fifteen were hospitalized, and seven died. The most likely source was droppings from wild birds in two inadequately protected water storage towers. Although admittedly a rare occurrence, this episode points out the possibility of such infections and the need for surveillance.

Several outbreaks of dysentery, a severe form of diarrhea, have been traced to the bacterial species *Shigella*. Two outbreaks in 1992 were associated with swimming in lakes. Two outbreaks occurred in 1994 from untreated well water. A total of 437 and 266 people, respectively, were affected by these outbreaks. These numbers are lower than those reported for 1986, when more than 2,000 people became ill in four unrelated outbreaks related to water-borne sources.

Protozoa

Protozoa are one-celled parasites that are members of the kingdom *Protista*. They act like tiny animals and were once called animacules. They are larger than bacteria, but a microscope is still needed to see them. They exist ubiquitously in nature and in water. Some that exist in water give rise to intestinal problems if ingested.

Because fewer protozoa than bacteria are needed to give rise to diarrhea, and because protozoa are more resistant than bacteria to the disinfectants that are used in municipal water systems, the total number of protozoan water-borne infections outnumbers the number of bacterial infections. Our knowledge of water-borne protozoan infections is relatively new, however, and less is known with certainty about their biology and behavior than is known about bacterial infections. Many of these infectious agents have come to the public's attention only in the last decade. In this section we describe the three most common water-borne types of protozoan infection.

Cryptosporidium

Cryptosporidium parvum is a protozoan that lives in the intestines of animals. Through the ingestion of water contaminated by animal feces, or through touching animals or infected people and not properly washing one's hands, it can reach the intestines of humans. There it gives rise to a disease called cryptosporidiosis. This illness is characterized by body aches, gas, bloating, weight loss, vomiting, low-grade fever, and especially watery diarrhea. Healthy people are usually sick for several days (rarely more than ten to fourteen days), after which they recover uneventfully. Some may continue to shed the organism in their stools for several months, but they are usually asymptomatic.

There is considerable evidence that low levels of *Cryptosporidium* occur endemically in most water supplies in this country. In one study, *Cryptosporidium* was present in 65 to 97 percent of the surface waters in very low concentrations. In another study of people with gastrointestinal complaints (diarrhea, nausea, and cramping), the prevalence of *Cryptosporidium* in stool specimens was 1 to 4 percent.

There have been at least six well-documented outbreaks of cryptosporidiosis from drinking water. The first was in Texas in 1984, the next in Georgia in 1987. In the latter outbreak 17,000 people became sick. The largest outbreak, however, was in 1993 in Milwaukee, Wisconsin, when more than 400,000 people became sick. Other outbreaks have occurred from bathing in swimming pools and water amusement parks. What makes all of these outbreaks unique is that they were contracted from surface water that had been properly treated and disinfected with chlorine. In the summer of 1998, the city of Sydney, Australia also had a waterborne outbreak of *Cryptosporidium.*

Although cryptosporidiosis is unpleasant, it is almost always self-limiting, with complete recovery. However, there is an exception to this rule—people with impaired immune systems (medically they are said to be immune compromised or immunocompromised). The immune system, which consists of white blood cells and antibodies, is the body's internal system of protection against infection from the outside world. People whose immune systems are impaired may not recover from infection by *Cryptosporidium* and may develop chronic diarrhea.

At the time in the 1980s that *Cryptosporidium* became recognized as a water-borne pathogen (source of illness), the acquired immune deficiency syndrome (AIDS) epidemic began to become prevalent. Characteristic of AIDS is a weakened immune system that cannot effectively ward off infections. Simultaneous with the beginning of the AIDS epidemic, the number of people receiving chemotherapy for cancer and the number of people receiving immune suppression therapy for organ transplantation also increased. All these groups of immunocompromised individuals are at significantly increased risk of developing a chronic form of cryptosporidiosis, with life-long diarrhea and risk of dehydration. The quality of life of these people is poor. It is estimated that approximately 2 percent of people whose cases of AIDS are reported to the Centers for Disease Control and Prevention have cryptosporidiosis as their AIDS-defining illness. In studies of hospitalized AIDS patients, the prevalence of cryptosporidiosis is 10 to 20 percent.

The reason for the persistence and tenacity of *Cryptosporidium* is that during part of its existence (the cyst or oocyst phase) it is encased in a hard shell that makes it impervious to antibiotics. It is also for this reason that it is able to withstand conventional techniques of water disinfection, such as chlorination. From a public health perspective, this means that other treatment methods are needed to safeguard the welfare of the community if this parasite is not to give rise to further epidemics.

However, there are uncertainties with regard to *Cryptosporidium* that hamper such public health efforts. It is not clear what minimal dose, or *inoculum*, of this protozoan is necessary to seed the intestinal tract of a person to give rise to the disease cryptosporidiosis. Is it just one or two organisms, or is it hundreds? This issue is of importance, since the answer to this question would determine to what extent drinking water needs to be treated so as to avoid infection. One study has suggested that ingestion of as few as 20 oocysts can lead to illness in 20 percent of people and that ingestion of 100 oocysts leads to illness in 40 percent.

Cryptosporidium oocysts have been detected in treated water in this country. The numbers are generally low, ranging from 0.001 to 0.006 oocysts per liter of water. Much of this variation depends on the source of the water. Water originating in rivers and lakes has higher concentrations than water originating underground, where the risk of contamination from animal waste is lower. During outbreaks of cryptosporidiosis, cyst concentrations have been found to be much higher. In the Georgia outbreak, for example, the cyst levels were as high as 2.2 oocysts per liter of water. In the Milwaukee outbreak, concentrations were believed to have reached 0.16 oocysts per liter.

To date several methods have been found to be effective in preventing water-borne cryptosporidiosis. These methods should be considered by anyone with a weakened immune system who has doubt regarding the reliability of his or her drinking water. The first method is filtration, that is, mechanically removing the oocysts. To be sure that all oocysts have been removed, water should be filtered through an "absolute" filter of 1 micron or less; that is, the diameter of the hole through which the water passes should be less than 1 millionth of a meter. Such filters are available for home use, and they should be certified as "NSF # 53" approved. (The National Sanitation Foundation or NSF is a company that tests and approves filters. The code for the filter that is used is 53. See Chapter 10 for details on water purification). Ozonation, reverse osmosis, and distillation (see Chapter 10 for details) are other methods of water purification, but are expensive and cumbersome. However, bottled waters are commercially available that have been ozonized or distilled (see Chapter 9). Perhaps the easiest and most economical method of purification is boiling, after which the water should be refrigerated in a clean bottle. We should mention that the purchase of any type of bottled water is no guarantee that the water will be *Cryptosporidium* free. As we discuss in Chapter 9, bottled water is subject to contamination. To ensure that bottled water is sterile and *Cryptosporidium* free, one must be sure to purchase water that is distilled or ozonized.

In December 1998 it was announced that *Cryptosporidium* in drinking water will be monitored by the government for the first time. The details of the implementation are not yet finalized. What is clear is that any person with a weakened immune system should not drink water from lakes or rivers without proper treatment. It is also fairly certain that recreational swimming also poses a risk.

Giardia

Similar to cryptosporidiosis, the illness giardiasis is caused by a one-celled protozoan, *Giardia lamblia.* Also like cryptosporidiosis, giardiasis is characterized by intestinal symptoms of bloating, gas, and diarrhea that last one week to ten days. Rarely does it go on to a chronic phase. The parasite *Giardia lamblia* is encased in an outer shell that allows it to exist outside the body for prolonged periods of time. This makes it resistant to disinfection. Like *Cryptosporidium*, the *Giardia* parasite gains access to the intestinal tract through infected water, as well as person-to-person contact with unclean hands (as in day care centers) or through homosexual contact. Unlike cryptosporidiosis, however, giardiasis can be combatted by effective antibiotic treatment. The most frequently prescribed medicine is metronidazole (brand name Flagyl). Also used are furazolidine (Furoxone) and paromomycin (Humatin).

Giardiasis occurs worldwide. It ranks among the top 20 percent of infectious diseases that cause the greatest amount of morbidity in Africa, Asia, and Latin America. It is estimated that in these parts of the world more than 2 million people per year become infected with it. In the United States, giardiasis is the most frequently identified cause of diarrhea associated with drinking water. Since 1965 more than 80 water-borne outbreaks have been reported, affecting more than 20,000 people. Of those cases of diarrhea reported to the CDC for which a causative agent was found, *Giardia* was identified nearly two-thirds of the time.

In this country people at highest risk of water-borne giardiasis are those who live in cities whose water supplies are derived from surface water that is not filtered, hikers and campers who drink untreated water, and international travelers. More than 20 million Americans live in cities whose water supplies do not routinely undergo filtration. At greatest risk are those whose water supplies come from clear running streams in mountainous regions that are exposed to animal droppings. It is for this reason that hikers and campers are also easily contaminated. With regard to travel, visitors to underdeveloped countries are known to be at high risk. The water in St. Petersburg, Russia, is known to be infested with *Giardia* as well.

The number of cysts of *Giardia* necessary to induce disease in a person is not known with certainty, but it is believed that only ten cysts are needed. Since the *Giardia* parasite replicates in the intestine every twelve hours once it is released from its shell, the ingestion of one parasite can, in theory, give rise to more than 1 million oocysts within ten days. This is more than enough to cause illness.

The efficacy of water treatment for preventing the spread of giardiasis is uncertain. Disinfection with chlorine at doses generally used in many water treatment plants (0.1 to 0.5 milligram per liter) is ineffective for destroying *Giardia.* There is as yet no agreed-upon level of chlorine that is safe and that can reliably kill *Giardia.* At times of *Giardia* outbreaks municipalities can increase the chlorine content of water, but the water is then often unpalatable. The pore sizes of most sand filters (the filters most commonly used by municipal water suppliers) are not sufficiently small to completely remove *Giardia* from the water.

Therefore, avoidance of infection depends to a large extent on maintaining clean source water and pretreating the water with coagulating particles so as to make the oocysts precipitate out with other organic particles. Unfortunately, this is a costly process that many small water suppliers cannot afford. If one is concerned about contracting giardiasis, the easiest thing to do is to boil the water. This is the advice one hears when there is a "boil only" notice.

When boiling is not convenient or possible, as for campers, several other options are available. A portable water filter with a diameter of no larger than six microns (six millionths of a meter) can be used. Some of these filters also contain iodide-impregnated resins that add extra protection. For those without filters another alternative is chemical disinfection with iodine or chlorine. High concentrations are recommended, up to 8 milligrams per liter. Either of these chemicals must be in the water for up to eight hours to be completely effective. Ascorbic acid (vitamin C) can be added to the water later on to remove some of the medicinelike taste that results from such treatment.

Cyclospora

Recently one more name has been added to the list of parasites that can cause diarrhea through ingestion from contaminated drinking water: *Cyclospora cayetanensis.* Little is yet known about this microbe, though it is a close cousin of *Cryptosporidium.* If untreated, it gives rise to a protracted illness lasting for weeks, and often relapsing. Fortunately the antibiotic trimethoprim-sulfamethoxazole, commonly known by its brand names, Bactrim and Septra, is an effective treatment.

The largest outbreak of cyclosporiasis to date was indirectly linked to water. This outbreak occurred in 1996. More than 1,500 cases were reported over a wide area of North America, almost all east of the Mississippi. Such an outbreak attracted the attention of public health officials, who immediately began to compare those who became ill with those from similar locations who did not become ill. It soon became clear that the individuals who had become ill had one thing in common: they had eaten raspberries! Unique to these raspberries was the fact that they had been imported from Guatemala.

When health officials visited the raspberry plantations in Guatemala, a potential source of contamination became evident. As raspberries are grown off the ground at shoulder height, surface contamination is unlikely. But when fertilizers and fungicides are used to spray the fruit, these chemicals are diluted with surface water from streams on the farms. The raspberries are washed after picking with water from the same sources. As anyone who has eaten raspberries knows, the surface of a raspberry is a labyrinth of crevices, an opportune place for viable parasites to remain dormant until ingested.

In other parts of the world, cyclosporiasis has also been found to be associated with water. Nepal and Peru are two places where this has been documented. In the United States *Cyclospora* has also been found to be water borne. In 1990 21 hospital workers who attended a party in a Chicago hospital became ill from it. The stirring of stagnant water in a storage tank after repair of a water pump was believed to have caused the contamination. In another report from Utah, contaminated sewage that seeped into a home was implicated as the source of infection.

To date there is insufficient information to allow one to say to what extent *Cyclospora* presents a public health problem. Continued observation will be necessary.

Viruses

Viruses are the smallest infectious organisms known to man. They easily pass through the smallest filters and vary in size from 15 to 300 nanometers (billionths of a meter). They are incapable of reproduction outside of a living cell. They can be rodlike, spherical, or polyhedral. They consist of a coat of protein enclosing a central core of nucleic acids called DNA or RNA. Through attachment to and control of their host's DNA they are able to replicate. Viruses are the organisms responsible for the colds from which we suffer, but they also cause such devastating diseases as hepatitis, AIDS, and rabies.

Until the 1970s, diagnostic tests were able to identify only bacteria and protozoa as the causative agents of infectious diarrhea. In a significant proportion of cases no causative agent could be identified. In 1972, using an electron microscope, researchers found a virus in stool specimens of people with epidemic diarrhea. The electron microscope can magnify objects many thousands of times smaller than can be seen with a regular microscope. The disease causing agent was called the Norwalk virus. In 1978 another viral agent was found using electron microscopy. This one was called the rotavirus. Since that time many other diarrhea-producing viruses have been identified.

The Norwalk virus is a representative of a class of viruses called small round structured viruses (SRSVs). In the United States, diarrheal illness due to SRSVs most commonly occurs among people of school age or older. Rotaviruses, on the other hand, are the most common causes of diarrhea among small children. In this country approximately 3.5 million cases of rotavirus-induced diarrhea occur each year, with an estimated 75 to 125 deaths. Approximately one-third of pediatric diarrheal admissions to the hospital in this country are due to rotavirus. Worldwide there are upwards of 140 million cases, with nearly one million deaths. Vaccines to combat this virus became available in 1998.

Most epidemics of these viruses are propagated through person-to-person transmission or are due to ingestion of filter-feeding shellfish (clams and oysters). Occasionally infected water from municipal water supplies, well water, stream water, lake water, commercial ice, or recreational pool water may serve as a source. Owing to the difficulty of diagnosing these outbreaks, their true extent is unclear. Many undoubtedly go unreported or unrecognized. It is known, however, that of thirty water-borne diarrheal outbreaks reported to the CDC in 1993 and 1994, no causative agent could be identified in five (16.7 percent). It is likely that these were due to viruses. In all instances inadequate water treatment was found. In several instances untreated noncommunity well water was to blame, and in others inadequate chlorination was the factor. Generally, with adequate chlorination of drinking water, viral infections are prevented.

Pfiesteria

A new water-borne infection that has recently been described is *Pfiesteria piscicida*. It is an alga that has been reported to live along the eastern seaboard of the United States, from Maryland to the Gulf of Mexico. Almost all of the reports of *Pfiesteria* are recent, beginning in 1996 and 1997. As its name—"fish killing"—implies, it is implicated in numerous fish kills, especially in inland rivers and estuaries. En masse, these algae are called the red tide. Their mechanism of fish killing is to release a poison that sickens fish and eventually kills them. Fishermen and residents of inland rivers where *Pfiesteria* has appeared report being sickened by the smell of the toxin that these microorganisms emit. There are also reports of temporary short-term memory loss and neurologic deficits.

To date *Pfiesteria* is not a problem that has impacted drinking water. Indeed, it is currently a problem only for fishermen whose livelihood has been impacted. Yet if *Pfiesteria* spreads to more inland estuaries, a problem may develop in the future. It is too early to say, and not enough is known. Many state and federal agencies are aware of the potential for problems and are therefore studying this microorganism. It is believed that one reason that *Pfiesteria* began to multiply was that waste water rich in farm fertilizers and nitrates was being dumped into coastal waters. Thus a human intervention related to water processing may have upset the balance of nature and triggered an environmental epidemic. Only time will tell.

Conclusion

In this chapter we have seen how drinking water can serve as a medium for the spread of infectious diseases. The type of infectious disease that is spread is in part a reflection of the society in which it occurs. In the developing world, water-borne infections such as cholera, typhoid, and infectious diarrhea persist due to poverty and poor sanitation. In the industrialized world, illnesses such as cryptosporidiosis that stem from newly discovered pathogens or medical therapies are of more concern. This concern has spurred regulators to reevaluate the need for further control of water sources and water processing.

It should be noted that despite all the known and possible infectious agents that may be present in drinking water and its sources, the drinking water in the United States is of very high purity and safety.

5

Drinking Water
and the Risk
of Cancer

The term *cancer* conjures up many negative
and painful emotions. Chronic suffering, distressful treat-
ments, futility, and fear are associated with this group of illnesses. So strong is the emotional impact of cancer that a
"war" on it was declared in the 1970s. The battle continues
to this day, with large government outlays for cancer
research and prevention.

A discussion of drinking water and the risk of cancer
concerns itself with the identification of chemicals and contaminants that appear in a sufficient number of water supplies at a high enough concentration to pose a potential
increased risk of cancer in the general public. Most of the
government regulations surrounding drinking water crafted
in the 1960s and 1970s were developed with cancer prevention in mind.

To approach a discussion of cancer and drinking water,
we have divided this chapter into several sections. First, we

define cancer and describe the mechanisms that are believed to underlie its development. Second, we describe how substances are determined to cause cancer. Third, we detail the contaminants found in drinking water that are believed to be related to cancer production. We also discuss how these substances reach drinking water.

What Is Cancer, and What Causes It?

Cancer is the uncontrolled growth and spread of cells. Cancer cells divide abnormally because they contain damaged or abnormal DNA, the nucleic acid molecules that dictate how a cell functions and how it will divide. Every cell contains proteins that attempt to repair damaged DNA. When repair proteins cannot repair damaged DNA, the affected cell(s) most often cannot divide and the damage is not propagated to a new generation of cells. On the other hand, if the DNA that controls the process of cell division is damaged and not repaired, affected cells may divide more readily than normal cells. These cells become cancerous.

DNA is damaged on a regular basis. Random events occurring during a cell's life account for much of this damage. Exposure to chemicals and radiation account for the rest. Chemicals or radioactive substances that are capable of damaging DNA and therefore giving rise to cancer are referred to as carcinogens. For a substance to be considered a carcinogen, several different types of evidence must be evaluated. First, experiments are done on cells in the laboratory to test whether the substance damages DNA. These are often done on cells of bacteria. Second, experiments are done on living animals (generally mice and rats) to determine if exposure results in the development of malignant tumors. During these types of experiments, efforts are also made to determine the level and duration of exposure that lead to tumor formation. Many government agencies, academic centers, and businesses are involved in this type of research.

The last method of determining carcinogenicity would be performing studies in humans. Since intentional exposure of individuals to known or suspected carcinogens is unethical, this type of research is not done. Rather, indirect proof is sought through analysis of unintentional or naturally occurring exposures. This is usually done in one of two ways. The first is conducting large ecological or environmental studies in which one group of individuals is or has been exposed to a certain factor and another group has not. It is then determined whether the group that was exposed to the factor(s) develops a higher rate of cancer than the group that was not exposed.

For example, if one wants to know if smoking is related to lung cancer, one can determine the rates of lung cancer in a group of smokers and in a group of nonsmokers. If the rate among smokers is higher than the rate among nonsmokers, it is possible to infer that smoking is related to lung cancer. One study alone is not sufficient to deduce such a conclusion, but if numerous studies yield similar results the case becomes stronger. To strengthen the case further, if a "dose response curve" can be demonstrated—that is, if the more one smokes, the greater the risk, or the less one smokes, the lower the risk—the argument for a relation between smoking and lung cancer is even stronger. Finally, if other factors known to cause lung cancer, such as exposure to asbestos, are taken into account and the relationship between smoking and lung cancer persists, the relationship between the putative carcinogen and the cancer becomes most convincing. Although this form of evidence is inferential and does not directly prove that a factor causes cancer, the weight of evidence becomes stronger and stronger until it is finally accepted as conclusive.

The second method of studying possible carcinogens in people is the case control method. With this method one examines a group of people with and without a specific illness and tries to deduce, looking back (retrospectively), if there were any differences between one group and the other in terms of past exposures. If it is found that subjects with lung cancer were many times more likely to have smoked than individuals who did not develop lung cancer, it can be deduced that smoking is related to lung cancer. Again, one study does not make for a definite association. Repeated studies and increasing weight of evidence are needed for a convincing argument to be made and accepted.

The problem with these approaches is that they are not as exact or reproducible as laboratory conditions. Often in the course of a study one is asked to recall exposure factors from many years ago. Imagine attempting to recall what one ate or drank ten or twenty years ago or how many cigarettes one smoked as a teenager! Such recollections are imprecise and fraught with error. Similarly, studying a large community, such as a city or a state, for exposure to a certain factor over ten to twenty years is also imprecise. Since most cancers take years and decades to develop, this is a very real problem. The population is constantly changing, levels of exposure change, and newer exposures and factors are appearing. Therefore, epidemiological and ecological studies are open to serious question and are often viewed with skepticism.

Many of these issues plague the arguments regarding the potential carcinogenicity of several contaminants found in drinking water. Moreover, many of the chemicals in water that are believed to cause cancer do not cause large increases in risk when compared to well-established carcinogens such as cigarette smoke or inhaled asbestos. The increased risk of these well-known carcinogens is maybe 10 to 100 times higher in the exposed compared to the unexposed. On the other hand, most water-borne carcinogens have increased risks one and a half to three times higher in the exposed than in the unexposed. These relatively small increases in risk associated with these substances often can be explained by other risk factors, making definitive conclusions regarding their carcinogenicity inherently less certain. For that reason, many issues related to the carcinogenicity of water-borne substances are unresolved. The process of deciding whether a substance is or is not a carcinogen is therefore a long and complex one.

The International Association of Research on Cancer (IARC) is an international agency based in Lyon, France that was set up in 1965 by the World Health Organization and is sponsored by sixteen governments. Its mission is to coordinate and conduct research on the causes of human cancer. The IARC regularly convenes panels of experts who assign potential carcinogens to one of five groups based on the existing evidence:

- Group 1—The agent (mixture) is carcinogenic to humans.
- Group 2A—The agent (mixture) is probably carcinogenic to humans.
- Group 2B—The agent (mixture) is possibly carcinogenic to humans.
- Group 3—The agent (mixture) is not classifiable as to its carcinogenicity in humans.
- Group 4—The agent (mixture) is probably not carcinogenic in humans.

The IARC panels evaluate evidence from human studies and animal studies and other basic science data before assigning each agent to one of the five groups. Evidence is periodically reassessed and updated. The results of these panels are published as a series of monographs by the IARC and are available on the world wide web (see the web site www.IARC.fr). In Appendix 3 is a list of the contaminants found in drinking water that are believed to be carcinogenic. Their IARC scores appear next to their names.

Types of Water-borne Carcinogens

A discussion of the types of potential carcinogens that are present in drinking water can be organized in several ways. For our purposes we classify carcinogens based on their point of entry into the water supply. The first group is those that enter through the natural geology from which the water originates. The second group is those that are added to the water supply from contamination and human commerce at the site of the water's source. The last group is those that appear during the processing of water.

Potential Carcinogens Originating from the Local Geology

This section deals with arsenic, asbestos, and radionuclides.

Arsenic. Arsenic is a metal used in many industrial processes. Several decades ago it polluted many of the nation's rivers and its air. Owing to improved metal refining techniques, improved air quality control, and better control of industrial spillage, arsenic exposure from water and air pollution is now much less of a risk than it was several decades ago. Water-borne arsenic is now derived mostly from geologic contamination of source waters. This occurs in areas rich in volcanic sediment. Especially high concentrations are present in Taiwan, Argentina and Chile, Mexico, Bangladesh, and India. In the United States, areas of the southwest and Alaska have high natural levels.

Several studies have shown that long-term ingestion of water-derived arsenic at high concentrations can lead to several skin conditions. These include thickening of the skin, an increase in skin pigmentation, and skin tumors. Dose response curves indicate a relationship between the amount of arsenic exposure and the number of cases of skin cancer. This relationship has been very well established in Bangladesh, where arsenic poisoning from contaminated well water has affected millions of people. The dose response curves have led to the general acceptance of arsenic as a bona fide skin carcinogen.

More recently there have been several reports of a relationship between arsenic exposure and internal cancers. In ecological studies from Argentina, lung cancer rates were reported to be increased in regions with high water concentrations of arsenic. In Taiwan lung, kidney, bladder, and colon cancers were found to be increased in regions with high concentrations of arsenic. In this country premalignant changes of the bladder were found in people from Nevada who drank water high in arsenic. Recent laboratory studies show that arsenic can stimulate the expression of genes involved in cell proliferation. For these reasons the IARC now classifies arsenic as a group 1 carcinogen.

Although this evidence suggests a link between these cancers and water-borne arsenic, much remains uncertain. Many of the studies that demonstrate a relationship between arsenic and internal cancers have been done in communities with very high levels of arsenic in the water, often several hundred micrograms per liter. These levels are much higher than those found in almost all drinking water sources in this country. From these data it is not clear whether one can extrapolate that the lower the level of arsenic, the lower the risk of cancer, or whether there is a threshold concentration below which the risk of cancer from ingested arsenic disappears.

This issue is not a purely academic one. The current interim recommendation of the Environmental Protection Agency (EPA) for water concentration of arsenic is 50 micrograms per liter or less. This level is based on risk calculations of a lifetime probability of developing skin tumors and does not take into account the newer information on the possibility of internal tumors. The studies from Taiwan suggest, however, that at this level there is an increased risk of internal cancer. The rates of bladder cancer, for example, are calculated to be especially elevated, at an incidence rate of two to five extra cases per one thousand people exposed. This is a level far in excess of levels deemed acceptable for exposure to potential environmental carcinogens.

As a consequence of this finding, Congress passed an amendment to the Safe Drinking Water Act in 1996 to study whether water arsenic concentrations should be further lowered. It has been proposed that the acceptable level be only 2 micrograms per liter. A committee's report is due in the year 2000. To date the World Health Organization (WHO) has accepted a provisional level of 10 micrograms per liter, and Canada has adopted a level of 25 micrograms per liter with the intention of eventually lowering the level further.

In this country the water concentration of arsenic for 98 percent of the population is less than 10 micrograms per liter. If the acceptable level of arsenic in drinking water is lowered in the future, it should not prove to be a hardship for most communities to meet this goal. On the other hand, approximately 2 percent of the U.S. population uses drinking water with higher concentrations of arsenic. In some cases that level may be as high as 50 to 100 micrograms per liter. Most such water is from small local water sources. It is for the people using this water that the new regulations may impose a change. For now, if one's water is derived from a private well or small community source, it would be prudent to check its arsenic concentrations.

Asbestos. Asbestos, a silica mineral, is a known lung carcinogen. Occupational studies of workers done nearly fifty years ago showed that those exposed to silica dust were many times more likely to later develop certain types of lung cancer than those who were not exposed. It was also found that stomach cancers were common among those exposed to asbestos, possibly due to the swallowing of lung mucus that contained asbestos that then came in contact with the stomach lining.

The carcinogenic effect of asbestos in drinking water is unclear. Asbestos is widely distributed through the environment and can reach drinking water sources through weathering of local geological sites, through pollution of water sources from mining and refining of iron, or through erosion of asbestos-based water pipes and conduits. Asbestos exists in a fiber form, and the physical characteristics of the fibers (length, width, and diameter) are as important to assessing its carcinogenic potential as is its concentration (number of fibers per liter). Studies of water-borne asbestos and cancer have yielded conflicting results. In studies from San Francisco, associations between naturally occurring asbestos in water with stomach, esophagus, and lung cancer were found. In Quebec, Canada, lung and stomach cancer associations were reported. In Norway stomach cancer rates were found to be elevated. On the other hand, in Puget Sound in Washington State and in another Canadian survey, no relationships with cancer were detected. In Duluth, Minnesota, whose waters came from Lake Superior, inconsistent results were found. From 1955 to 1973 the waters of Lake Superior had high asbestos levels owing to industrial pollution. In some cases cancer rates were elevated, but the relationships were inconsistent.

Overall, there is no compelling evidence that naturally occurring asbestos in water serves as a vector for carcinogenesis. Nonetheless, because asbestos is a known pulmonary carcinogen, its level in drinking water is regulated by the Safe Drinking Water Act and it is classified as a Group 1 carcinogen by the IARC.

Radionuclides. The relationship between ionizing radiation and cancer is well known. Survivors of atomic explosions and atomic accidents are at a high risk of cancer. Less clear is whether the amount of naturally occurring radiation that appears in drinking water can also play a role in carcinogenesis. Trace amounts of radium, uranium, and radon appear in drinking water supplies all over the world. The concentrations vary. Surface waters such as lakes and rivers have the lowest concentrations, whereas private wells have the highest concentrations.

Several ecological studies suggest a very weak link between cancer and exposure to water containing radionuclides. In a study from Florida, comparing rates of cancer in areas of high water radionuclide concentrations to areas with low levels of radionuclides in water, leukemia rates were found to be higher in the areas of high water radionuclide concentrations. A study in Iowa suggested a mildly increased leukemia rate. In North Carolina the incidence of childhood cancers of the blood and brain was found to be higher in areas with high radon levels than in areas with low radon. All these studies, however, are open to criticism with regard to methods of measurement and exposure, as well as the inability to take numerous other factors into account. For these reasons it is believed that in this country very few cases of cancer can be directly attributed to radionuclide contamination of water.

Potential Carcinogens Originating from Contamination of Water Sources by Human Pollution and Commerce

Many chemicals produced by humans find their way into the drinking water supply, either intentionally or inadvertently. Chemicals that are intentionally dumped into rivers and lakes are the most obvious examples. Insecticides and fungicides that are introduced into the soil are another example. Yet even chemicals that are not in direct contact with the ground or with water sources ultimately reach water sources through evaporation and precipitation. Because many of these chemicals are long lived, they are able to accumulate in water at relatively high concentrations. In this section we discuss the two main types of human water pollutants: nitrates and organic chemicals.

Nitrates. Agricultural runoff containing nitrogen-rich insecticides, herbicides, rodenticides, fertilizers, and animal manure has led to a marked increase in groundwater nitrate levels. The annual fertilizer production alone in the United States is more than 10 million tons of nitrogen equivalents. Several studies have shown that groundwater in areas near agricultural sites may have from three to sixty times the levels of nitrates as areas that are not agricultural.

Cancer concerns related to water-derived nitrates come from several studies that have shown that ingested nitrates can interact with the lining of the stomach to form N-nitrosamines, chemicals that are known to be potent carcinogens. Studies have also shown that the amount of nitrate intake has an effect on the amount of N-nitrosamine production. Despite these concerns, the evidence of an increased cancer risk from exposure to water-derived nitrates is weak and inconclusive. Part of the problem is with the inherent difficulty of measuring community exposure to nitrates over a long period of time, which is compounded by the long latency periods needed for the development of tumors.

In France no connection between cancer rates and water nitrate concentrations has been found. On the other hand, in Spain stomach cancer levels were found to be elevated in areas with high water nitrate concentrations. A similar finding was reported from England. Two other studies from England, however, did not demonstrate such an association. In the United States several studies from Illinois found no cancer connection, whereas a study from Nebraska suggested an increased rate of lymph node cancer (lymphoma). Likewise, case control studies of patients with stomach cancer have given conflicting results.

In the United States the maximum water contaminant level for nitrates is 10 milligrams per liter. Almost all large community water supplies fall within this range. Private wells, on the other hand, often have higher levels. This is especially so in agricultural sections of the country. In a study by the U.S. Geological Survey, almost 10 percent of randomly chosen wells were above the safe limit. Although the risk of cancer from these high concentrations remains uncertain, it would be prudent for those who rely on private or small water systems to have their water checked for nitrates.

Organic Chemicals. The post–World War II industrialization of the U.S. and world economies has led to the introduction of tens of thousands of new chemicals into industry, agriculture, commerce, and domestic life. Although these have made our lives easier, more efficient, and more comfortable, they have also produced a problem for the environment. Many chemicals are durable, and their disposal has become a problem. As an attempted solution, many chemicals have been dumped into waterways. Others have been put into landfills. Either way, many or most of these substances have gained access to drinking water supplies. Most often these substances accumulate locally, but if they gain access to underground aquifers the potential for widespread dissemination is increased. This is especially so for fertilizers and pesticides that are directly introduced into the soil and seep downward. Other chemicals that decompose and evaporate may regain entry into the water system in the form of precipitation.

The chemicals that cause the greatest problems are those referred to as organic chemicals. They contain carbon atoms. Those that decompose or evaporate are called volatile organics. These include such compounds as benzene, carbon tetrachloride, and vinyl chloride. Exposure to these could occur through ingestion, inhalation, or skin contact. Other organic compounds are more stable. These include such compounds as acrylamide, chlordane, dioxin, and polychlorinated biphenyls (PCBs). Exposure to these would be through ingestion. Many of these chemicals have chlorine atoms attached to them and are part of a large family of organic molecules called organochlorines (see below).

In Appendix 3 is a list of water contaminants that are believed to be carcinogenic. As can be seen, the majority of these substances are organic chemicals. Of these chemicals benzene, vinyl chloride, and dioxin are IARC Group 1 carcinogens. Seven other chemicals are rated by IARC as Group 2A substances, probable carcinogens. Eleven others are in IARC Group 2B, possible carcinogens. Several others are not classified.

Relatively few studies have examined the cancer consequences of these contaminants. Those that have been done have been hampered by the problems that are present in most ecological and epidemiological studies. To overcome these deficiencies, several studies have been done in areas that were on or near toxic dump sites. In these sites concentrations of organic chemical contaminants were many times higher than those commonly found in the environment. By studying such areas it was hoped that a clear trend or association could be detected.

One such study was done in Love Canal in New York State, near Niagara Falls. That region served as a toxic waste dump site from 1947 to 1952. Later on, housing was built near the site. An assessment of cancer rates in that area for the period from 1955 to 1977 showed an excess of lung cancer. Whether this increased rate was due to contaminated water was unclear, however, since many of the contaminants were able to bubble up to the surface, then evaporate into the air, and they could have been inhaled. In a study from Woburn, Massachusetts, a cluster of individuals with childhood leukemia was reported where community waters were contaminated with organic chemicals. On the other hand, in another study of residents near a toxic waste site in Riverside County, California, no excessive cancer rates were reported.

In 1989 a nationwide study of cancer rates in 339 counties with toxic waste dump sites was reported. Significantly elevated rates were found for lung, bladder, stomach, and colon cancers. In New Jersey leukemia rates were found to be elevated in 27 townships with high concentrations of volatile organics in the drinking supply. In Pennsylvania bladder cancer rates were elevated in areas with contaminated water supplies.

These studies show that elevated levels of chemicals in drinking water *may* be associated with an increased cancer risk. However, because of the inherent difficulties associated with these studies, they do not prove that each of the substances listed in Appendix 3 are carcinogenic. Nonetheless, prudence dictates that the levels of these substances be kept as low as possible. As before, people who rely on small water distribution systems or private wells are at greatest risk of having these contaminants present in their water. Such people should strongly consider having their water tested for contaminants.

Carcinogens Originating from Water Processing

In this section we discuss the most controversial and interesting aspect of water contamination and cancer risk—that associated with chlorine. It is ironic that a substance that has contributed as few others have to the public's welfare by eliminating many infectious pathogens from water should now be at the center of a controversy regarding its potential to cause cancer.

As a chemical element, chlorine forms strong electron bonds with carbon atoms. Therefore, carbon-chlorine (organochlorine) compounds are stable and resistant to degradation. This allows them to persist and accumulate in the environment, including soil and water. Because these compounds are lipophilic—that is, easily accessible to fat—they are able to accumulate in the fat of animals and humans when ingested in food and water.

Organochlorine products were first identified in drinking water in the mid-1970s. They result from the combination of chlorine, added during the processing of drinking water, with naturally occurring organic compounds found in water due to decomposed vegetation and human pollution. Most of these compounds have not been identified to date, and those that have been identified have not been studied extensively. The major factor determining the concentration of these compounds is the level of organic compounds in the source waters. Because there are more organic compounds and industrial wastes in surface waters than in groundwaters, surface waters have higher concentrations of these chemicals than do groundwaters.

Among the organochlorines that have been studied, the most commonly occurring are the trihalomethanes (THMs) and haloacetates. THMs are carbon compounds to which three chlorine atoms are attached. Chloroform is an example of a THM. In some cases bromine atoms are attached in place of chloride atoms (these substances are called brominated THMs) if the bromide level in the water is elevated. THMs occur in a concentration of 30 to 300 micrograms per liter in surface waters and one to 10 micrograms per liter in groundwaters. In studies of mice, chronic exposure to THMs and chloroform have been implicated in liver and kidney tumor formation. Brominated THMs cause DNA mutations in bacteria, as well as kidney and colon cancers in mice. Among the haloacetates are compounds such as trichloroacetic acid (TCA) and dichloroacetate (DCA). Trihaloacetates have an acetate group associated with three chlorine atoms. Assays in laboratory animals show that these substances induce liver tumors.

The first studies in humans to examine whether chlorine byproducts increased cancer rates were geographic-type studies. They compared rates and types of cancers in areas with and without chlorinated waters. Cancers of the colon, rectum, and bladder were found to be increased in areas with chlorinated water. Later, case control studies were done that appeared to confirm these results. In one study the relative risk of having bladder cancer was 1.6 to 1.8 times higher in those who were exposed to chlorinated water for thirty years or more than in those who were never exposed to chlorinated water. One large study suggested that approximately 5,000 cases of bladder cancer and 8,000 cases of rectal cancer per year could be attributed to the consumption of chlorinated water in this country. More recently a report from

Denmark purports to show a relationship between organo-chlorine exposure and breast cancer.

The strongest evidence to date of a relationship between the ingestion of chlorinated water and cancer is for bladder cancer. The results are less consistent for rectal cancer. Owing to the weaknesses inherent in geographic and case-control studies, there is still a great deal of uncertainty as to how strong the relationship is between cancer and chlorinated water consumption.

Because of this uncertainty, there has arisen a debate in the environmental and industrial communities regarding the continued use of chlorine for water decontamination. The environmental group Greenpeace has argued that, because organochlorines do not occur naturally in the environment and because industrial organochlorines are potentially harmful, the use of chlorine should be banned. Further, they say, the number of organochlorines is so great and their types are so numerous and hard to classify that it will be impossible to ever be sure that chlorine byproducts are safe. They believe such a ban should include both industrial use of chlorine and its use for water treatment. They point to the fact that there are now alternatives for water treatment that are not associated with the production of organochlorines.

Industry representatives counter by saying that the bad substances among the organochlorine compounds have been identified and removed. They say that there is no compelling argument to date that organochlorine byproducts from water treatment are harmful. Their concentrations in water are low, and laboratory testing uses higher doses than those found in drinking water. Besides, even if chlorination byproducts were harmful at their current concentrations, the amount of good they do by disinfecting drinking water of microorganisms far outweighs their possible adverse effects by many, many orders of magnitude. The newer methods of water purification (such as ozonation) are expensive and unproven on a large scale. Even if it were possible to sterilize drinking water through alternate means at a treatment plant, the conduits that carry the water to taps would still require chlorination to maintain water purity during distribution. Finally, chlorine is so widespread throughout industry and water that any attempt to find a substitute would be very expensive.

Where the truth lies in this debate is unknown. Most probably somewhere in the middle. Chlorine products may be associated with increased cancer risk. As we will show in the next chapter, organochlorines are also implicated in problems of procreation. On the other hand, there is no denying that one of the greatest steps in public health has been the chlorination of municipal waters and the reduction of water-borne infectious diseases. What is clear and agreed upon by all is that more research is needed and that the solution to these problems will slowly emerge over time.

6

Drinking Water, Estrogens, and Fertility

In the early 1990s considerable public interest was aroused by a medical report from Denmark. In that report researchers purported to have found evidence that in western Europe the average sperm count had been steadily falling over the last fifty years, and that the quality of the sperm was also reduced. The decline in sperm count was not insignificant, as it was up to 40 percent in some countries. In addition, the report indicated that there was an increase in the number of cases of undescended testicles (in which the testicles remain in the abdomen and do not go down into the scrotum, the sac where the testes usually reside), hypospadias (a malformation of the penis in which the opening for urine is on the underside of the penis, not at the tip), and tumors of the testicles, which are usually rare.

The authors of the Denmark study put forward a hypothesis to explain these findings. According to them, the decline in sperm count and the increase in the number of genital and testicular problems were due to an overexposure of the male fetus during gestation (growth in the uterus) to too many estrogenlike substances that the mother, in turn, had obtained from the environment. This overabundance of estrogenlike substances during the formative period of the male fetus caused the delayed formation of the sex hormone–producing cells in the testicles, which later in adulthood led to a decreased sperm count. The overabundance of estrogen-like substances during pregnancy also had an effect on the formation of the male fetus's sex organs.

Before proceeding, we will explain these concepts and put them into perspective. Later we will show how they relate to drinking water.

Estrogens

What are estrogens? They are hormones, or chemicals, that are normally produced in the ovaries of women. They stimulate the growth of the womb (uterus) and breasts, and are responsible for other secondary sex characteristics in women. Estrogens are also produced in men, but in much smaller quantities.

Estrogens are released into the bloodstream by the ovaries. They then circulate throughout the body, searching out targets, or receptors, on the surfaces of cells or inside the cells to which they can bind. When they attach to the receptors on or in the cells, they trigger the cells to undergo various changes. For example, when estrogens attach to the receptors of the cells of the uterus, they initiate the process that leads to menstruation. In young girls going through puberty, estrogens stimulate breast cells to grow so that in the coming years they reach a mature size.

Estrogens circulate in the blood at very low levels. They exist at such low levels that their concentrations are measured in parts per trillion—the equivalent of less than a grain of sugar in an Olympic-sized swimming pool! Despite this, they are able to do their work efficiently owing to their high potency and ability to "lock into" specific receptors in specific tissues, such as the uterus and breasts.

Normally a fetus is exposed to its mother's estrogens during gestation (the process of development during pregnancy). The levels of estrogens to which it is exposed are not high. This is owing to the presence in the blood of proteins, called binding proteins, which bind the estrogens. They allow only a small percentage of the estrogens to be free to attach to receptors, while the majority remain bound to the proteins to be released as the need arises. In such a way the effect of estrogens is closely controlled. In such an environment, a fetus develops normally.

However, when this natural order is upset and a fetus is exposed to higher than normal doses of estrogens, abnormalities in development occur. How do we know this? Unfortunately, an inadvertent human experiment taught us. In the 1950s and 1960s more than five million women were given a medicine called diethylstilbestrol (DES) during pregnancy in the hope of preventing spontaneous miscarriages. DES is a powerful estrogenlike medicine that can interact with the cell receptors for estrogen, even though it is not chemically similar to estrogen. The thinking then was that the more estrogen or estrogenlike substance that was available to the mother, the more likely she was to carry her pregnancy to term. For that reason DES was widely prescribed. Because it was not carried by the binding proteins in the blood the way natural estrogens are carried, DES was free to attach to estrogen receptors at a rate higher than that of natural estrogens.

Later it was learned that high levels of estrogen or estrogenlike substances during pregnancy were not beneficial.

Indeed, women who took DES had higher rates of spontaneous abortions than women who had not taken DES.
Moreover, there were adverse effects on some of the children
of women who were so treated. Male children were noted to
have an increased rate of testicular abnormalities, as well as
lower sperm counts as adults. Female children were found to
have anatomic abnormalities of their genital tracts that led
to infertility. A few unfortunate women even developed a
rare cancer of the vagina in their early adult years that
required surgical removal of the vagina. What this ill-fated
medical episode taught us was that exposure to an abnormal
level of estrogen or estrogenlike compound at a critical junc-
ture during gestation could have an impact on the fetus.
Moreover, that impact was not necessarily an immediate
effect, but could become noticeable many years later. In
other words, an apparently minor change in hormone levels
during a critical stage of gestation could leave a lasting
imprint on the genetic expression of an individual and alter
him or her for life. A tightly controlled hormone environment during critical stages of fetal development was crucial
for normal development.

Estrogenlike Substances

The other lesson learned from DES was that a chemical substance that did not chemically resemble estrogen
was able to interact with the estrogen receptor and act like
estrogen. DES has a totally different chemical structure from
that of natural estrogens. Despite this, DES was able to interact with the estrogen receptor and mimic estrogen's effects.
This suggests that the estrogen receptor is not as highly specific as other hormone receptors and is actually quite oblig-
ing or "promiscuous." This leaves open the possibility that
other substances can interact with the estrogen receptor and

cause problems similar to those caused by DES if they were to come in contact with the fetus during critical stages of development. Evidence that this is indeed the case began to appear soon after the DES scare.

In the 1960s it became appreciated that many of the 50,000 to 100,000 new chemicals that had been developed in the post–World War II industrial boom were harmful to the health of the environment and to the health of the individual. These were chemicals used in daily living. Among them were pesticides, of which DDT was the best known; polychlorinated biphenyls (PCBs), which were found in hydraulic fluids, adhesives, and flame retardants; dioxins, which were byproducts of incinerators, paper processing, steel mills, and motor vehicles; and alkylphenol polyethoxylates (APEs), which were used in paints, detergents, cosmetics, and herbicides.

The harm or "toxicity" of these chemicals was assessed in terms of immediate or intermediate consequences. For example, if inhalation or ingestion of a compound led to an immediate untoward effect, such as shortness of breath or birth defects in fetuses, the substance was deemed dangerous. Standards were then developed to determine the highest permissible level of exposure below which no untoward immediate effects would occur. This form of protection was quickly instituted and is still used today. In terms of intermediate harm, prevention efforts focused on the cancer-promoting effects of these chemicals. This measure of toxicity was adopted on account of the numerous tumors that were found in animals and fish living in areas that were polluted with these substances. This measure was also adopted on account of the unique and frightening concern that cancer evoked (and continues to evoke) in all of us.

In the 1960s and 1970s the first evidence that chemicals may have even longer-term toxicity came from observational studies of wildlife. Animal groups that had hitherto shown normal procreative function and parental skills were noted to have impaired fertility, ambiguous genitalia, and deficient parenting behavior. In Lake Apopka, Florida, alligators had a significant drop in the number of juveniles and an increased number of juveniles with ambiguous genitalia. Fish living in polluted streams, lakes, and ocean waters had higher than expected rates of genital abnormalities. Birds living along the Pacific coast had skewed ratios of female to male hatchlings and did not exhibit normal parenting behavior. Among mammals, Florida panthers and minks living along the Great Lakes exhibited a number of developmental and reproductive abnormalities, including low sperm counts and poor sperm function.

How these disparate events came to occur and how they were related to one another remained unclear for many years. Only through painstaking observation and the persistence of a small cadre of environmental scientists and toxicologists was a cohesive explanation gradually developed and tested. It became clear that many of the same chemicals that caused short- and intermediate-term toxicities were also acting as estrogenlike substances and were interfering with normal fetal sexual development. How could this happen? The answer lies in several characteristics of these chemicals.

One of the defining characteristics of the many chemicals that were produced after World War II is their persistence. Many resist breakdown by natural means and many persist for years or even decades. Therefore, vast amounts accumulate in the environment. Another characteristic of these substances is that they are fat soluble and accumulate in animal fat. Every animal on the food chain that comes in contact with a source of food that had previously been in contact with these chemicals, concentrates these chemicals in its body. This is the process of bioaccumulation. For example, crayfish living in waters polluted by a factory upstream concentrate chemicals in their bodies at levels many hundreds of times higher than in the surrounding waters by eating zoo- and phytoplankton that are contaminated with chemicals. The crayfish, in turn, are eaten by larger fish, which further concentrate the chemicals in their body fat at levels another thousand times higher than those seen in the crayfish. The fish, in turn, are eaten by small animals that are then eaten by larger predators such as alligators or eagles. With each step up the food chain, the concentration of chemicals in the body fat increases. Humans, at the top of the food chain, are thus exposed to the highest animal tissue concentrations of these substances. These chemicals, in turn, are concentrated in the human body's fat. Through such a progressive process humans may attain levels of concentration of chemicals many thousands to millions of times higher than those in the environment. Because these chemicals are highly stable, the human body has few ways of ridding itself of them.

Yet another characteristic of these chemicals is that they can interact with the estrogen receptor. Their interactions are not as strong as those of natural estrogens, but their sheer amount and concentration in the body are sufficient to cause an effect. Although their absolute concentrations in fat is low, measured in parts per million and parts per billion, one should remember that the body's natural stores of estrogens are measured in parts per trillion. The order of magnitude of the concentration of these chemicals is therefore many times higher in the body than that of naturally occurring estrogens, and through mass action they can interact with estrogen receptors. This is not surprising, in hindsight, since the chemical structures of many of these substances are similar to that of DES.

Perhaps, however, the greatest breakthrough in our understanding of estrogenlike compounds was the realization that the toxicity of these chemicals is greatest during fetal development. Prior to that, toxicological research had been largely confined to the study of adults and had not considered the effects on fetal development. The new research suggested that the most toxic effects of estrogenlike compounds are on the developing fetus. In many elegant experiments, minimal changes in the concentrations of estrogen or estrogenlike substances during fetal development led to alterations in subsequent sexual behavior and preferences, masculine or feminine behaviors, sex organ development and sperm counts, and the ability and desire to procreate. Changes in the levels of concentration of these substances, therefore, imprinted themselves on genetic destiny and altered that destiny during early fetal life.

It was only natural, then, to extrapolate from these findings that many of the abnormalities noted in male children and adults were due to exposure to estrogenlike chemicals during human gestation. During pregnancy the chemicals that were stored in the mother's body fat were released into her bloodstream as fat deposits were broken down to provide the growing fetus with food. As a consequence, throughout pregnancy the fetus was exposed to high levels of estrogenlike compounds and the potential for endocrine system disruption was present.

Water and Estrogenlike Chemicals

Returning to the study presented at the beginning of this chapter, it is now believed that much of the infertility and increased sex organ anomalies seen in humans is similar to that seen in animals. Just as animals appear to be affected by estrogenlike compounds during fetal development, humans also appear to be so predisposed. At this point this theory is unproven, but the DES episode shows that it is possible. The EPA and other government agencies are now actively studying this subject.

The question now arises as to how these findings are relevant to the topic of this book—drinking water. The answer is that water—the same water that we rely on for drinking—serves as a conduit for the dissemination of these chemicals. Between the end of World War II and the early 1980s, hundreds of billions of pounds of chemicals made by humans were released into the environment. In 1989, for example, more than five billion pounds of pesticides were produced worldwide. In the United States alone, 2.2 billion pounds were used on nearly one million farms and 70 million households. Billions of pounds of other chemicals made by humans were further added to the environment either intentionally or as waste products of manufacturing. In 1992 more than 400 billion pounds of synthetic chemicals were produced in the United States alone.

As would be expected, much of this industrial output has found its way into the water supply. Some of it reaches water through direct dumping of factory waste into rivers, estuaries, and lakes. Other indirect sources, however, also contribute. Many of the chemicals are incinerated or evaporate, entering the air and atmosphere. They are then dispersed worldwide. During precipitation they are reintroduced into the water supply in the form of rain. This helps explain how chemicals produced in one site can appear at another distant site. Indeed, high levels of chemicals such as DDT and PCBs have been found in the Arctic.

Yet another source of water pollution is the seeping of chemicals into the soil and eventually into underground streams and aquifers. For example, pesticides are intentionally introduced into land. They slowly work downward into the soil. High concentrations are also found in landfills, which eventually contaminate nearby water sources. All these potential sources of contamination eventually lead to pollution of water sources. They, in turn, act as the medium by which fish and other animals, upon which we rely for our food, become contaminated.

An example of the pervasiveness of this problem comes from several surveys. In California in the late 1970s, more than 2,500 wells were found to be contaminated with levels of various organic chemicals up to five times higher than recommended. Further sampling of wells of large California water systems for organic contaminants showed that 18 percent had some contamination and that 5.6 percent exceeded one or more state-mandated standards. Almost half of the contaminated wells were in the Los Angeles area, suggesting that industrialization and urbanization played a role in water contamination. A similar sampling of smaller, more rural California water suppliers showed that almost 8 percent of wells had signs of contamination and that 2.1 percent exceeded state regulations for at least one contaminant. Similar surveys done in New Jersey showed that 6 percent of wells exceeded government standards for at least one compound.

As a result of these findings, the EPA has set up strict standards to regulate the levels of chemicals in the drinking water supply. Nonetheless, these standards cannot control for the amounts of chemicals that we ingest in the form of food—animal tissues in which chemicals have accumulated or grains, fruits, and vegetables that have been sprayed with pesticides and fungicides. The exposure is ubiquitous. If future studies show that estrogenlike compounds are indeed responsible, or partially responsible, for some of the infertility problems and the genital abnormalities in humans, seeking a solution will be difficult and costly.

More recently, there has been further speculation that the estrogenlike effects of chemicals may be responsible for other medical problems. In this country women's lifetime risk of breast cancer has been steadily increasing over the past several decades. It is estimated that a woman has a lifetime chance of one in nine of developing this dreaded disease. One of the few factors known to predispose a woman to breast cancer is her lifetime exposure to estrogens; the more estrogen exposure, the higher the risk. Scientists are now exploring whether estrogenlike compounds may play a role in this epidemic.

Animal studies suggest that fetal exposure to estrogens stimulates the growth of the prostate gland or of uterine cells. Could widespread exposure to estrogenlike chemicals explain the recent rise in prostate cancer seen in men in this country? Could such exposure also explain the recent increased prevalence of endometriosis in young women? Endometriosis is a disorder in which uterine tissue appears in the abdominal cavity outside of the uterus. It is believed to be due to excess estrogen exposure. It affects more than 5 million women in this country, causes significant pain, and is a leading cause of infertility.

Even more intriguing than these questions are questions related to the possible effects of estrogenlike compounds on the brain. Several studies have shown that children of women who ate fish regularly during pregnancy—and presumably had higher ingested levels of estrogenlike chemicals—had higher rates of learning disabilities, attention deficit disorder, and hyperactivity. Could these findings be related?

In April 1998 a report was published by the World Resource Institute in Washington, D.C. It indicated that over the past several decades there had been a shift in the proportion of male to female births in the industrialized countries of Europe and North America. Fewer males were being born. This report is similar to reports in the 1960s of birds living along the Pacific Coast that had skewed female to male birthing rates. Could it be that the same thing is now happening to humans, quietly and imperceptibly? Time will tell, but it should be borne in mind that we, like the birds, are part of nature and are not immune to the detrimental effects that our pollution has on the environment.

These issues are still controversial and unproven, but they point to the possibility that chemical exposure may have far-ranging effects on our lives if occurring during certain critical junctures.

Conclusion

There is a historical note of irony in the story of estrogenlike chemical compounds as they relate to our drinking water sources. Early this century it was learned that many epidemic illnesses were spread by flying insects in watery, marshy areas. Among these were malaria, dengue, and yellow fever, which was spread by the bite of a mosquito; sleeping sickness, which was spread by a fly; and typhus, which was spread by lice. With the introduction of the pesticide DDT during World War II, large areas were "debugged" and pesticidal epidemics were prevented. DDT was looked on as a wonder chemical and as the means of saving the lives of millions.

In a twist of fate, DDT and other pesticides have now been banned because of their effect on procreation among animals despite their proven ability to rid areas of disease-carrying insects and prevent disease. The very chemicals that cleared up water-related infectious diseases have come to endanger our health in other ways. One problem has been substituted for another. This fact should remind us that maintaining the safety of our water supplies is an ongoing process and that vigilance will always be necessary.

7

The Effects of Drinking Water's Mineral Content on Health

In this country drinking water is derived mainly from two types of sources: from sources on the earth's surface, such as rivers, streams, and reservoirs—called surface water—and sources from beneath the earth's surface, such as wells and springs—called groundwater. The rocks, soil, and local geology through which these waters travel add minerals and metals to water's content. The term *water hardness* refers to the concentration of these substances in natural water. Among these substances are calcium, magnesium, strontium, iron, and manganese, as well as a large number of trace elements such as fluoride, zinc, selenium, cadmium, lead, copper, and chromium. Because the most common of these elements are calcium and magnesium, *hardness* is often defined as the sum of the concentrations of these two elements or, more simply, as the concentration of calcium.

Water hardness can be detected in a number of ways in our daily lives. Clogged pipes and boilers caused by the precipitation of minerals may lead to reduced water flow, the need for costly repairs, and increased water heating costs. Many of the soaps and detergents used for laundering may have their sudsing and cleaning abilities compromised due to interaction with the water's calcium.

In areas where the water is hard or very hard, local water utility companies "soften" the water through a number of methods that precipitate out the minerals and metals before the water reaches the consumer. The resulting water hardness varies from community to community. Water with fewer than 75 milligrams per 100 cubic centimeters (mg/deciliter or mg/dl) or fewer than 60 parts per million (ppm) of calcium is considered soft. Water with 75 to 150 mg/dl or 61 to 120 ppm of calcium is considered moderately hard, and that with calcium concentrations above these levels is considered hard. A survey of treated water from the public water supplies in the one hundred largest cities in the United States showed that in more than 90 percent of the cities the calcium concentration was less than 50 mg/dl. Therefore, if one drinks water from the tap, in all likelihood the water is soft.

Recently the health effects of drinking water's mineral content have been examined. Several studies have suggested that minerals in drinking water are beneficial to well-being, and may help prevent or forestall several chronic diseases that become more common with aging. The interest in this subject is considerable, since small changes in water hardness can be accomplished at a low cost and potentially have a large impact on health at the community level.

In this chapter we discuss the effects of water hardness on cardiovascular disease and osteoporosis. Both are chronic diseases that occur late in life and may be delayed by preventive measures earlier in life. Water's mineral content and kidney stone formation are also discussed. Finally, we discuss the addition of fluoride to drinking water and its effect on dental health.

Water Hardness and Cardiovascular Disease

Beginning in the late 1950s and early 1960s, several geographic studies were conducted to determine why there were wide variances in the rates of heart attacks, strokes, and high blood pressure (collectively called cardiovascular disease) across many parts of the United States and the world. These studies called attention to a possible inverse relationship between the degree of water hardness and a decreased risk of cardiovascular disease; that is, the possibility that the higher the mineral content of the water in a given area, the lower was the risk of cardiovascular illness in that area.

Since then, nearly one hundred studies have been reported examining the relationship of drinking water hardness to cardiovascular disease. Studies done in large population areas, such as states and provinces, have supported the possibility that increased water hardness is associated with a decreased risk of heart attacks and strokes. The decrease in risk of such disease has been estimated to be as high as 10 percent, though most studies suggest that it is much lower. Studies done in smaller population areas such as cities or counties, on the other hand, do not support a reduction in such risk. This lack of consensus is not surprising since the effect of water hardness on cardiovascular disease does not appear to be a large one, and such an effect can be detected only in studies involving very large numbers of people.

One way that drinking water might protect against cardiovascular disease is through replenishing the body's stores of an essential element that protects the heart and blood vessels. In this regard, magnesium has become a widely discussed candidate. Magnesium is used by the body in many of its day-to-day functions. Nerve and muscle cells use it to keep their membrane sheaths (coverings) healthy, thereby maintaining their ability to conduct electrical signals, especially in the muscles of the heart. With a low level of magnesium, the cells' membranes no longer function well electrically. As a result, the nerve and muscle cells may become excitable and erratic in function. Low magnesium levels can lead to irregular heartbeat and rhythm disturbances, and in the blood vessels to spasms and reduced blood flow to critical organs. In such cases heart attacks and sudden death can occur.

The daily recommended dietary intake of magnesium is 6 milligrams per kilogram (mg/kg) of body weight per day. For the average adult man of 70 kilograms (154 pounds), that is approximately 350 milligrams; for an average woman of 60 kilograms (132 pounds), it is 280 milligrams. Drinking water is estimated to account for only about 2 to 4 percent of a person's daily consumption of magnesium, based on an average concentration of 6 milligrams of magnesium per liter (a little more than a quart) of water and an average consumption of 1.5 to 2 liters of water per day. Yet replenishment from drinking water may play an important role in people who may be mildly depleted of magnesium, such as those who take diuretics (water pills) for high blood pressure or heart disease and who lose excessive amounts of magnesium in the urine. The chemical form of magnesium that appears in water is also more readily absorbed by the body than that derived from food.

Studies in the United States, Canada, and Scandinavia have shown that the higher the magnesium level in water, the lower is the rate of cardiovascular death. It has been suggested that much of this decrease is due to a decreased rate of sudden cardiovascular death. Recently the U.S. Food and

Drug Administration (FDA) has contracted with the National Academy of Sciences to determine the benefits of magnesium-enriched water and to see whether this should become an issue of public health policy. One of the main things that the National Academy of Sciences will try to determine is whether the amount of magnesium in drinking water comprises enough of the magnesium humans need daily to have an effect on overall health.

The other element in hard water that may contribute to reduced heart attack and stroke rates is calcium. Several studies have suggested that the cardiovascular protective effect of calcium in drinking water is as strong as that of magnesium. As with magnesium, the effect of water-derived calcium is probably indirect. The normal daily ingestion of calcium in the United States is approximately 500 milligrams per day or less, depending on age and gender. Therefore, if we assume that an average person drinks 1.5 to 2 liters of water per day and that there are 50 milligrams of calcium per liter of water, approximately 15 percent to 20 percent of one's daily calcium requirements may be met through water ingestion. This quantity represents a relatively small part of daily calcium consumption and thus is unlikely to have a direct effect on the heart and blood vessels.

A mechanism by which water-derived calcium could reduce heart attacks and strokes would be its effect on high blood pressure. Studies have shown that a low level of calcium intake is associated with elevated blood pressure. Conversely, dietary calcium supplementation has been shown to lower blood pressure in individuals with high blood pressure. How this occurs is uncertain, but it is known that calcium exerts a strong effect on the contractility of muscle cells found in the blood vessel walls. Blood vessel muscles that contract may lead to high blood pressure owing to the greater force the heart has to exert to push the blood through the narrowed blood vessels. High blood pressure is a recognized risk factor that leads to increased rates of stroke and heart disease. Whether the amount of calcium derived from hard water is sufficient to decrease the contractility of the muscles in the blood vessel walls and thereby to lower blood pressure is uncertain. As in the case of magnesium, the results of further rigorous scientific studies are awaited.

We should note that there are several limitations to the studies that suggest that water hardness may be related to cardiovascular health. These studies are most often retrospective, and they usually demonstrate only a small effect. Therefore, it is not possible to rule out other competing explanations of the seemingly positive effects. Socioeconomic, cultural, demographic, nutritional, and genetic factors all could play a role in the results and yet are not or cannot be taken into account. It is also possible that what was being measured as "water hardness" in these studies was actually a surrogate for something else that was not measured. For example, magnesium and calcium frequently coexist in water. It is possible that the imputed effect of calcium is actually due to magnesium.

Nonetheless, there is enough evidence of the effects of water hardness on the cardiovascular system to warrant further scientific studies. Were such effects found to stand up to scientific scrutiny, an inexpensive public health measure that could impact the overall health of the community could be implemented.

Water Hardness and Osteoporosis

With the aging of the U.S. population, there has been an increasing emphasis on disease prevention as a means of forestalling the degenerative changes that are associated with advanced age. High on the list of these priorities, especially for women, is prevention of osteoporosis. Osteoporosis is broadly defined as a thinning of the bones. Osteoporotic bones have lost a large proportion of their mineral content and thereby their strength and resilience. This, in turn, leads to an increased risk of fractures following falls, especially fractures of the hip, as well as crush fractures of the spine that result in pain, shortened stature, and stooped posture.

Many medical studies have emphasized that oral calcium supplementation alone or in combination with female hormone replacement therapy, vitamin D, and other medications helps delay the onset of osteoporosis. Little is known, however, as to whether calcium derived from drinking water can also have an effect on delaying the development of osteoporosis. This may at first seem to be a point of little significance, since the vast majority of our daily calcium intake is derived from food, especially dairy products, green vegetables, and fish. On further thought, however, this consideration does have some merit.

For a woman, bone density reaches its maximum in the third and fourth decades of life. Thereafter it begins to slowly decline with aging. At the time of menopause the rate of bone loss accelerates, up to 2 to 3 percent per year for approximately ten years. Thereafter the rate of bone loss once again slows to 1 percent per year. By the age of sixty or sixty-five many women have lost up to 30 percent or more of their peak bone density. This is important since a woman's risk of osteoporosis and fracture is strongly related not only to how much bone density she loses during the aging process, but also to the starting point from which bone loss began. If a woman has a low peak bone density at a young age, that person has a lower density later in life after the effects of aging have set in and is more likely to suffer a fracture. Conversely, the woman with an initially high bone density will have a higher bone density later in life, and therefore a lower risk of fracture.

The average American woman has a relatively low calcium intake. It is estimated that after menopause women in the United States consume an average of only 475 to 575 milligrams of calcium per day. What is more, the ability to absorb calcium decreases with age. In recent years the amount of calcium that is needed to maintain proper calcium and bone balance has been reevaluated. Many researchers now believe that postmenopausal women need 1000 to 1500 milligrams of calcium per day to maintain healthy bones. If this is the case, high lifetime intakes of calcium from water rich in calcium may play an important role in achieving a high peak bone density. Likewise, high calcium content of water may also play a role in maintaining an adequate calcium balance during aging. A number of studies have suggested that high lifelong intakes of calcium may affect peak bone density and lead to fewer fractures. However, no studies have looked at the individual contribution of the calcium content of water to lifetime body calcium balance. It stands to reason, however, that in areas of borderline calcium intake and where other factors predisposing individuals to osteoporosis exist (e.g., diminished sunlight from prolonged winters) the calcium content of water may make a difference. Future studies will be needed to test this hypothesis.

Water Hardness and Kidney Stones

Small collections of calcified material that grow in the kidneys, called kidney stones, are a significant source of pain and suffering in the general population. As many as one in five hundred people in the United States has a kidney stone during the course of a year, and 1 to 5 percent of the U.S. population has them in their lifetime. In addition to causing pain, kidney stones also may lead to infections and to kidney damage.

Several factors are known to predispose individuals to the formation of kidney stones. First, many people have a genetic predisposition to developing stones. Stone formers tend to be clustered in families. Second, certain foods, such as animal proteins (meat) and dairy products, are associated with stone formation. Therefore, in times of famine or war when food is scarce, the number of kidney stones reported decreases. Finally, certain parts of the United States, such as the Southeast, have higher rates of stone formation than do other parts of the country. One explanation for this is the effect of the hot climate and the tendency of people in this area to become dehydrated more easily than in other parts of the country.

What leads to the formation of kidney stones is not entirely certain, but the following is the way most people believe they occur. When the kidney filters blood to remove excess salts, minerals, water, and contaminants, the resulting urine contains and concentrates these substances. Among these substances are sodium, potassium, calcium, magnesium, and phosphates. If urine is low on water relative to the concentration of these substances, these substances reach a point of saturation at which they no longer can be dissolved in the urine. In such a case, these substances precipitate out of the urine in the form of crystals, much as when there is too much sugar in one's coffee or tea. Stones form from these crystals.

It is known that calcium is the main component of more than 75 percent of kidney stones. The average amount of calcium excreted in the urine in twenty-four hours is usually more than 300 milligrams for a man and 250 milligrams for a woman. It would therefore be intuitive and logical to suppose that if one were to limit the amount of calcium in one's diet, including the amount derived from water, the incidence of kidney stones would go down. However, this does not seem to be the case! In areas of the United States with hard water, there are lower rates of kidney stone formation than in areas with low calcium concentrations in the water. This finding has been corroborated in other parts of the world, such as Scandinavia and Japan.

This finding is counterintuitive, and to date no good explanation has been offered. However, certain explanations are possible. For example, there may be a substance in hard water, other than calcium, that has an effect on the kidneys' filtering system whereby less calcium is excreted in the urine. Then, too, it is possible that a high amount of water-derived calcium may stimulate the intestines to excrete more calcium than the kidneys, thereby taking the burden off of the kidneys. Whatever the explanation, though, the best advice that someone with kidney stones could be given is to drink plenty of water (up to 2 liters) every day so as to maintain a dilute urine and not to worry about the water's mineral content.

Fluoride and Tooth Cavities (Dental Caries)

Compared to the lack of certainty regarding the effects of water hardness on cardiovascular illness, osteoporosis, and kidney stone formation, there is much more known regarding the effects of water fluoride content and the prevention of tooth decay or cavities (dental caries). Whereas naturally occurring fluoride contributes little to water hardness, its importance to public health cannot be overstated. Indeed, the knowledge gained about fluoride's effect on the prevention of tooth disease, and the translation of that knowledge into the deliberate addition of fluoride (fluoridation) to most public drinking water in this country, has been one of the major achievements of public health in the past fifty years. A brief history of how fluoride came to be recognized as a tooth decay preventive and how it came to be used will be helpful to the reader.

Early in this century, observations made in several regions of the United States and Europe showed that people living in certain well-defined areas, but not in adjacent areas, had discolored ("mottled") teeth. The teeth were often dark, and at times they were almost black. After further study it was determined that these discolorations were due to a substance in drinking water. Investigation at several laboratories revealed that fluoride was the agent responsible. In areas where there was endemic mottling of the teeth, fluoride levels in naturally occurring water were high (often greater than 2 milligrams per liter). In Europe, for example, the areas of high water concentrations of fluoride were found to be near volcanos. In areas where there was no mottling, levels of naturally occurring fluoride were usually normal (less than 1 milligram per liter).

During this time it was observed that children whose teeth were mottled (and who presumably had been exposed to high levels of fluoride) had much fewer dental cavities than children in the surrounding areas. In many cases the number of cavities appeared to be reduced by more than 60 percent. Studies were undertaken to confirm these observations and to find the water concentration of fluoride that would be protective against cavities and at the same time not lead to tooth discoloration. The conclusion reached by researchers was that a fluoride concentration of 1 milligram per liter was optimal. Above this level no added protection against cavities was found, but the risk of tooth mottling rose. Conversely, the more the concentration of fluoride was below the level of 1 milligram per liter, the greater was the risk of cavities with a low risk of mottling.

Following World War II, several controlled studies were undertaken to test whether the addition of fluoride to the public water supplies of cities with low fluoride levels would decrease the rate of cavities among children. On follow-up, cavity rates were found to be reduced by 50 to 65 percent compared to those for children in control cities in which the water had not been fluoridated. No adverse effects on tooth color were noted. These initial community-based experiments established fluoridation as a safe, effective, and inexpensive method of promoting the public's health. Since that time numerous studies all over the world have reconfirmed the benefits of water fluoridation and the lack of harmful side effects. There has been no increase in the number of reports of tooth mottling during the past forty years in either fluoridated or unfluoridated areas. (The only drawback to the use of fluoride in water is for patients who have kidney failure and who require a treatment called hemodialysis to cleanse their blood. Fluoride should not be used in the dialysis fluid since it may accumulate in the bones of dialysis patients. Specially prepared water is used for dialysis.)

In the 1990s it is estimated that approximately 60 percent of the U.S. population has been served by public drinking water that is fluoridated. More than 70 percent of cities with populations larger than 100,000 have their water fluoridated, and eight states have mandated that all large water supplies have supplemental fluoride added. On the other hand, many Americans ingest water that is naturally high in fluoride due to the local geology through which the water flows. Additional fluoride is then not needed. Such regions include many of the states from North Dakota to Texas and those along the Mexican border.

How fluoride exerts its effect is not entirely clear, but it probably binds to the enamel and dentin of teeth during the formative years of tooth development. Because fluoride deposits most rapidly during the period of tooth development, it is important that children be exposed to an adequate amount of fluoridated water during their formative years. Once the secondary teeth have formed and erupted, there is little further increase in tooth fluoride levels or strength. The Centers for Disease Control and Prevention have estimated that for every dollar spent on water fluoridation there is an estimated savings of $50 in dental care later in life.

For those parents who feed their small children bottled water and yet want the benefits of fluoride, there are now bottled waters with fluoride. These are available nationwide and are produced by such companies as McKesson, Culligan, and Perrier. A full list with telephone numbers is available on the web at www.bottledwater.org.

We should note that the amount of water consumed in different parts of the United States and the world varies. In those parts in which there is year-round warm weather, people tend to drink more water than in those parts with colder climates. Exposure to fluoride may therefore be increased. Accordingly, the levels of fluoride in the public waters are adjusted so that overexposure does not occur.

For more information regarding fluoridation, one can call one's local water utility. Information on fluoride levels in counties all over the United States are available through the Public Health Service, Centers for Disease Control and Prevention, National Center for Prevention Services, Division of Oral Health, 1600 Clifton Road, Atlanta, GA 30333.

Conclusion

In this chapter we have summarized how minerals in drinking water may have an effect on individual and public health. The associations of minerals with cardiovascular disease and osteoporosis are still tentative. Hopefully in the coming years scientific evidence will be deduced to support or negate the purported beneficial effects of magnesium and calcium in drinking water. If such effects are proven, the implications may be far reaching. Simple, inexpensive health measures could be undertaken to increase water hardness and, by implication, improve community-based cardiovascular and bone health efforts. Such steps, however, would have to be balanced against the increased costs of clogged pipes and home appliance malfunction due to deposition of minerals and scale. There is much stronger evidence of the benefits of fluoride, and fluoridation is widely practiced.

For the interested reader, Appendix 4 lists the mineral contents of the most commonly used bottled waters in the United States. Because the mineral content of municipal drinking water is not regulated, there are no national data available on individual cities. However, one can call one's local municipal government or regional EPA office to obtain information.

As a historical footnote, it is interesting to note how attitudes regarding the effect of water and its mineral content on health have come full circle. In antiquity mineral waters were considered important for health, but efficacious only when used for bathing. In the late sixteenth century the notion that drinking clean mineral water might have a medicinal purpose began to come into vogue. (At that time water was generally polluted and was often a source of illness.) The healing properties of many mineral springs began to be noticed by local residents, and experience showed that many illnesses could be cured by taking "the cure." Among the healing forces associated with these waters, "vitriol," a hypothetical substance believed to be present in water, was believed to help the body via its cleansing and astringent properties. Later, in the seventeenth and eighteenth centuries, with the rise of chemistry, efforts were made to assign cures to medical illnesses based on the actual mineral content of various water sources. Water cures became fashionable and widely practiced.

The rise of scientific medicine in the late nineteenth century and the twentieth century has led to a gradual decline in the use of water cures, though their use continues today in certain parts of Europe. The potential that drinking water's mineral contents can improve the public's health—this time based on sound principles of science—may yet lead to a resurgence of interest in the use of drinking water as a curative method.

8

Heavy Metal Content of Drinking Water and Its Effects on Health

Among all the different types of acute water-borne outbreaks that are reported to the Centers for Disease Control and Prevention (CDC)—infectious (including those caused by parasites, bacteria, and viruses), chemical (those caused by substances such as nitrates and PCBs from industrial runoffs), and metal—those caused by metals are the least common. The CDC estimates that there were only five or six reported cases or outbreaks of lead or copper poisoning in the United States in 1993 and 1994. This may in part reflect the low prevalence of these disorders, but it may also reflect the fact that many of these acute poisonings are ill defined and are not diagnosed. Then, too, many occur in private homes owing to corroded plumbing, and they may not come to public attention.

The focus of this chapter is on the effects of chronic excess metal exposure and intoxication from drinking water sources. These effects are subtle at onset, but in the long term are devastating. For the sake of brevity, we discuss four metals of the greatest interest and importance—lead, copper, aluminum, and mercury. Other metal contaminants are not discussed, but references concerning them are available in the bibliography.

Lead

Lead is a naturally occurring element that has been in use since the earliest recorded times. It has also been known for centuries that it may have a deleterious effect on health. The Roman author Pliny made mention of the "dangling . . . paralytic hands" of those who drank wine prepared in utensils of lead or water from pipes of lead that served as conduits for bringing water to homes or the public fountain. In all likelihood this is a reference to nerve paralysis arising from lead intoxication, a problem that was still present in nineteenth-century London. Because of lead's widespread use in many industrial and chemical processes, the level of lead exposure has risen, and with it has risen the potential for an adverse effect on the public's health.

The effects of lead exposure are most noticeable in the nervous system. The nervous system consists of the brain, the spinal cord, and the nerves that are present in the arms and legs. In children younger than three to five years old, the brain is most susceptible to the deleterious effects of lead. As a consequence, a child may suffer long-term neurological and behavioral disorders that carry over into adult life, such as hyperactivity, delayed physical and intellectual development, poor eye-hand coordination, and learning difficulties. Adults with chronic lead exposure may experience fatigue and impaired ability to concentrate.

Several other body organs may also be affected by chronic lead exposure, though less often than the brain. There may be a low blood count (anemia). In the bones and teeth there may be impaired cell growth and maturation, leading to short stature and poor dentition, respectively. Kidney function may also be impaired, as can testicular function, giving rise to low sperm counts. There may be wasting of peripheral nerves in the arms and legs, with loss of muscle tone and mass. There is also an association between elevated blood levels of lead and the presence of high blood pressure.

Because of the potentially long-term harmful impact of lead, in the past twenty years major efforts have been made to control exposure to it. The major sources of exposure are paint, auto exhaust, food, and water. The lead content of paint came under regulation in the late 1970s when it became clear that peeling and flaking paint from the walls of older buildings was a significant source of lead exposure. Small children would swallow paint chips and mouth items that had come in contact with leaded paint. As a result, paint was reformulated to have low levels or no lead. Although paint is now less of a problem than it used to be, many children remain at risk of this form of exposure. In cities where pre-1970s housing is still common, the risk is highest. Children in poor households now make up a disproportionate percentage of those with lead poisoning from peeling paint.

Also in the 1970s and 1980s, the content of lead in car gasoline came under stricter regulation. Car emissions were then a major source of lead exposure in the air, especially in urban settings where there is much traffic congestion. Following the introduction of low- and unleaded gasoline, inhalation of lead from gasoline emissions is no longer considered a public health problem.

Another source of lead exposure is food. Such exposure can occur in one of two ways. First, food itself may contain lead, especially vegetables and crops that grow in or near the ground and may absorb lead directly from the soil. Much atmospheric lead that is released from industrial plants finds its way back to the soil during precipitation. Also, farm machinery is not required to use unleaded gasoline, and the surrounding soil may absorb lead in concentrations higher than desired. Second, certain acidic foods have been found to absorb lead from lead solder in cans and ceramics and may serve as an additional source of lead in the diet. The lead content of solder in cans is now under regulation as well.

With the decrease of exposure to these sources, regulatory attention has been turned to lowering lead exposure from other sources of contamination, especially water. This is important since, as the levels used to define lead toxicity have dramatically fallen, it has become clear that neurological and behavioral impairments in lead-exposed children occur at levels once deemed safe. Levels as low as 10 micrograms per liter of blood are believed to be detrimental, whereas levels up to 45 micrograms per liter were deemed safe until recently (one microgram is one millionth of a gram; there are 454 grams in a pound). Therefore, further efforts at minimizing lead exposure are warranted especially in children.

There are two major sources of lead in drinking water. The first and less important is industrial pollution or contamination of soil by chemical pesticides and fungicides. For example, from 1987 to 1993 almost 144 million pounds of lead compounds were released from lead smelting industries to land and water, according to the Toxic Release Inventory (see selected EPA Federal Register documents, www.epa.gov/opptsfrs/home/histox2). The most common sites were Arizona and Missouri, with the largest releases to water occurring in Ohio. Fortunately, when lead is released to land it does not migrate to water, but remains in the soil. In water it sinks and binds to sediment, and very little is dissolved in the water. It does not accumulate in fish as does mercury, but it may do so in shellfish such as mussels.

The other major source of lead in drinking water is corrosion of the conduits and pipes through which the water is transported. It is this source that poses the greater risk to the public's health. Lead can be derived from the solder used to join copper pipes and from brass and chrome-plated brass faucets. In older buildings the pipes that connect the building to the main water service lines may also leach lead. (Perhaps 20 percent of main water service lines for public water supplies are made of lead.) Since 1986 the use of solder with more than 0.2 percent lead content has been banned, and the lead content of faucets and pipes has been restricted to 8 percent or less.

The Environmental Protection Agency (EPA) estimates that water-borne lead may make up 20 percent or more of a person's total lead exposure. This is important with regard to infants, who may be given water from the tap to drink or have their formulas and juices mixed with tap water. Likewise, in growing children water is an important source of lead exposure. In a 1994 study of water samples taken from 12,000 school and workplace water coolers, faucets, and ice makers, it was reported that 17 percent had lead concentrations of more than 15 micrograms per liter for a first-pass sample. The health effects of this exposure are unknown at this time.

To ensure that lead levels in drinking water are safe, the EPA (through the Safe Drinking Water Act) requires large water suppliers (those serving more than 50,000 people) to collect water samples from household taps twice a year. No more than 10 percent of the samples tested may have lead levels above 15 micrograms per liter. If the percentage of taps with high levels of lead exceeds 10 percent, a water supplier is expected to notify the public through the media and to take "corrosion control measures" so as to reduce the amount of lead in the water. Among these measures are changing the pH (acidity) of the water to prevent corrosion, and increasing the amount of calcium and phosphate in the water to promote the development of a protective coating of mineral sediment on the inside of the pipes. In the event that these measures are ineffective, the pipes themselves must be changed. The government estimates that monitoring the 79,000 community and nontransient noncommunity water systems that are required by law to be regulated costs approximately $40 million nationwide. This breaks down to ten cents per year per household for those serviced by large water systems and $3 per household per year for those serviced by smaller systems.

There are several things one can do as an individual to minimize one's exposure to lead from drinking water. The first is to flush water from a faucet before using it if the faucet has been unused for six or more hours. Water that has remained stagnant in a pipe tends to accumulate lead. Flushing for one or two minutes, or running one to two gallons, is generally sufficient. To avoid wasting water, water may be bottled for later use after flushing.

It is also best not to use hot water from a faucet for drinking or food preparation. Hot water dissolves lead more easily than cold water. Loose solder and debris within pipes and plumbing should also be removed by periodically removing faucet strainers from all taps and flushing the system for several minutes. One can contact one's building contractor or public utilities and water department to find out the materials used in the service line that delivers water to one's house or apartment and to obtain information about the community's drinking supply. If there is doubt one can call the Environmental Protection Agency's Drinking Water

Hotline at 800 426-4791 for more information or the Community Right-to-Know Hotline at 800 535-0202.

In the event that one finds lead levels to be elevated or is unsure whether they are elevated, there are two things that can be done. The first is to buy a home treatment device that filters water. Systems that use distillation or reverse osmosis (see Chapter 10 for details) are effective in removing lead. They require upkeep, however, and need replacement with time. If in doubt, the surest way of all to avoid excess water-borne lead is to purchase bottled water (see Chapter 9), which is required under government regulation to have safe lead levels.

Copper

Copper, like lead, finds its way into the water supply mostly through corrosion of pipes and conduits through which the water is transported. Rarely does copper reach any significant concentration in source waters unless there is direct dumping of copper into these waters. Most copper waste is dumped into landfills.

In 1992 the Centers for Disease Control and Prevention reported three outbreaks of gastrointestinal disease—i.e., nausea, vomiting, diarrhea, and abdominal pain—associated with high water levels of copper from newly installed copper plumbing systems. In 1993 and 1994, forty-three people were reported to have had gastrointestinal illness from elevated copper levels in tap water in a hotel due to the building's plumbing system, which caused prolonged stagnation of water. In the same year the American Association of Poison Control Centers reported 887 cases of copper poisoning. The true extent of the detrimental effects of elevated copper in the water supply is unclear.

The acute gastrointestinal symptoms caused by ingestion of water-borne copper are usually short lived. There are no suggestions that elevated levels of copper in water can lead to long-term health problems. A rare genetic disorder called Wilson's disease, which is due to the body's inability to get rid of copper that accumulates through absorption from the diet, has never been shown to be related to the copper content of water.

The maximal permissible level of copper in drinking water is 1.3 milligrams per liter. Any level above this requires action to be taken by the water company, much in the same manner that is done for elevated lead levels.

Aluminum

Aluminum is a commonly found element on the earth's surface. In the form of aluminosilicate, it accounts for 8 percent of the earth's crust by weight. It is not surprising, then, that as water courses through a region's geological formations aluminum becomes a naturally occurring substance in drinking water. Generally no more than 200 micrograms per liter are found in drinking water, though the concentrations are usually much lower. Therefore, if the average person drinks 1.5 liters of water per day, he or she takes in no more than 300 micrograms per day of aluminum from water. (Aluminum-containing chemicals, called alum or aluminum sulfate, that are used to trap dirt and large particles in surface waters in preparation for use as drinking water, settle out and do not add to the aluminum content of drinking water.) In contrast to these minuscule amounts, it is estimated that the average person ingests 10 to 20 milligrams of aluminum per day from the food that he or she eats.

It seems strange, then, that the small amounts of waterborne aluminum that a person drinks have been implicated as a possible cause of a major health problem that has recently gained considerable public attention—Alzheimer's disease. Alzheimer's disease is a degenerative disease of the brain occurring most often in older individuals. It is characterized by progressive loss of mental capacities, including memory and the ability to reason and to care for oneself. Ultimately the victim of Alzheimer's disease dies after a number of years of slow, progressive deterioration. The disease is reported to affect up to 5 to 10 percent of the population over the age of sixty-five years. It is a major public health problem, as it is a significant source of medical expenditure in the care of the elderly and puts a great emotional burden on the family that cares for a person suffering from this disorder.

Were one to do a pathological examination of the brain of an Alzheimer's patient, one would find "plaques" and "neurofibrillary tangles" in certain regions of the brain called the hippocampus and the cortex. These plaques and tangles are believed to be degenerating nerve cells, the same nerve cells that are responsible for cognition and reasoning. Interest in a possible association of these plaques with aluminum was first raised in the 1960s when a group of researchers was able to produce plaquelike lesions in the brains of experimental animals by injecting small amounts of aluminum salts into them. This work, in turn, led to several pathology studies of human brains from patients who had died of Alzheimer's disease. These studies showed an increase in the amount of aluminum in the brains of those with Alzheimer's disease compared to the brains of individuals who died of other neurological causes.

Further support for a role for aluminum in the onset and propagation of Alzheimer's disease came in the 1970s, when hemodialysis became a widely used method for treating patients with kidney failure. Hemodialysis is the process whereby blood is withdrawn from and returned to the body to cleanse it of the impurities that accumulate in it when the kidneys do not work. In order to prevent the accumulation of one of these substances, phosphorus, aluminum was added to the fluid that was used to cleanse the blood. Aluminum-containing antacids were also given by mouth to bind phosphorus in the intestines to prevent its absorption into the bloodstream. Soon reports began to appear of an increased prevalence of slurred speech, cognitive impairment, and death—symptoms similar to those of Alzheimer's disease—among dialysis patients. This neurologic deterioration, called dementia, was traced back to the high doses of aluminum that the patients had been receiving during treatment. In the absence of kidney function, the aluminum accumulated in the body and led to brain dysfunction. Although it was later discovered that the brains of patients with dialysis dementia differed somewhat from those with Alzheimer's disease, it was clear that aluminum could be associated with dementia.

These findings have led to several ecological and epidemiological studies that have examined the effects of aluminum exposure from drinking water on cognitive impairment and the incidence of Alzheimer's disease. In a study from Norway in 1986, a significant correlation was reported between deaths from dementia and the concentration of aluminum in drinking water. Later a 1989 study from England purported to show that the number of cases of "probable" Alzheimer's disease was 50 percent higher in districts whose water supplies had high aluminum levels than in districts with low water concentrations of aluminum. In 1991 a study from Ontario, Canada, demonstrated a relationship between hospital discharges of patients with Alzheimer's disease and aluminum concentrations in the drinking water. A second longitudinal study that same year from Ontario also showed that high concentrations of water-borne aluminum were more likely to be associated with higher rates of cognitive impairments in the population than in districts with low water concentrations of aluminum. Finally, in a recent study from the same group in Canada it was estimated that almost a quarter of the cases of Alzheimer's disease in the Ontario region could be causally linked to excess aluminum exposure from drinking water.

Although these studies are intriguing, they have several serious drawbacks that make their acceptance difficult. First, the cause of Alzheimer's disease and other forms of dementia are poorly understood. Therefore, it is hard to know reliably where aluminum may lie in the chain of events that lead up to dementia. It is possible, for example, that the deposition of aluminum in the brains of patients with Alzheimer's disease may be a result of Alzheimer's disease and not a cause of Alzheimer's disease. Second, many of the ecological studies have been hampered by an inability to take into account many extraneous confounding factors, such as people's moving in and out of communities with varying water aluminum concentrations and the possibility that other metals may be present in the water along with aluminum. Then, too, in many studies the diagnosis of Alzheimer's disease was based on imprecise criteria and not on brain specimens. Finally, proof is needed to demonstrate that the aluminum dissolved in water is more readily absorbed by the body and of greater toxicity than that found in food. The majority of food-derived aluminum is known not to be absorbed by the intestine, but there have been no studies to show that aluminum that is dissolved in water is any more easily absorbed than food-borne aluminum. Until these issues can be resolved through scientifically rigorous studies, the issue of water-derived aluminum toxicity remains an uncertainty. In the future this issue will no doubt be more carefully studied. For the moment, owing to uncertainty, aluminum concentrations in drinking water are not under enforceable government regulation.

Mercury

Compared to the metals lead, copper, and aluminum, which can dissolve in water and can come directly in contact with the body through drinking, mercury is rarely present at any significant level in water. Its effect on health is indirect, mainly through the fish whose meat we eat. Since the fish we eat come from the same aquatic systems from which we derive our drinking water, it is appropriate to briefly discuss mercury in the context of drinking water.

The deleterious effects of mercury exposure have been known since the industrial revolution of the nineteenth century. These adverse effects, affecting mainly the nervous system (the brain, spinal cord, and nerves in the arms and legs), are incurred through the inhalation of fumes that are the byproducts of factory processes. Only in this century, however, was it learned that mercury can also cause damage when water serves as a medium of exposure.

In the 1950s a chemical plant in Minimata Bay, Japan, discharged large amounts of mercury into the surrounding waters. Fishermen and their families who ate the fish from that area were soon stricken with severe neurological diseases such as paralysis and loss of feeling in the limbs. Some even died. Especially affected were babies whose mothers ate the contaminated fish while pregnant. Many of these children had developmental delays.

This outbreak illustrated a property of mercury—that it can bioaccumulate. Mercury is discharged as an inorganic form (unattached to a carbon atom) into water and returns to humans in an organic form (attached to a carbon atom) in concentrations many thousands to millions of times higher than in water. This is so since plankton are able to efficiently absorb mercury from the waters in which they live and to concentrate it in their tissues. Fish, in turn, eat the plankton and further concentrate the mercury in their tissues. (The reader may recall from Chapter 6 that many chemicals made by humans, such as dioxins and PCBs, also undergo such a concentrating process.) Humans, who are at the top of the food chain, are thus exposed to the highest concentrations of mercury when they eat fish.

Mercury reaches the earth's waters in one of two ways. The first is from direct dumping of industrial pollutants into rivers and estuaries or through the contamination of groundwaters in areas of toxic waste disposal. The other source of mercury contamination in water is through the release of mercury vapor from the earth's crust and volcanos and through the release of fumes from incinerators, mining, and the burning of fossil fuels. These emissions, in turn, return to water in the form of precipitation. Since they may linger in the atmosphere for months, they are distributed globally. No source of water escapes some contamination. This explains why mercury may be found in wide-ranging ocean fish and freshwater fish in sites where there are no industrial discharges of waste products. In areas where acid rain precipitates the mercury from the atmosphere, it appears that the acid leads to an increase in the attachment of mercury to organic compounds.

In the United States there have been no reported cases of mercury poisoning from fish caught in lakes with high mercury levels. However, populations that are dependent on fish or sea mammals (e.g., whales) as a major source of protein (for example, the Inuits in northern Canada) have at times developed blood levels of mercury that overlap the lowest levels of those who were poisoned in Minimata, Japan. To protect against excess consumption of mercury, regulatory agencies such as the EPA have set "safe" limits on the amount of mercury that can be found in edible tissues of fish. Knowing these levels and the dietary habits of people allows for a fair estimation of total dietary mercury intake. In addition, the EPA has set up a National Listing of Fish Consumption Advisories for all the states and territories. In it is a database containing information on bodies of water whose waters are believed to be polluted with high levels of mercury (as well as PCBs, chlordane, dioxins, and DDT) and in which fishing is not recommended. As of 1995, 1,740 water bodies were included. This represents 15 percent of the nation's total lake acreage and 4 percent of the nation's rivers. It includes the entire region of the Great Lakes and its connecting waters and a large percentage of the nation's coastal waters.

Conclusion

We have shown that high levels of heavy metals in drinking water can be associated with long-term neurologic problems. The effects of lead are well known, and lead concentration in water is therefore regulated. The effects of aluminum are only now being appreciated. Future research should help clarify whether aluminum concentrations in drinking water will also be under regulation.

In Appendix 1 are listed the allowable levels of the various metals in drinking water that we have discussed in this chapter, as well as several others that we have not discussed.

Part Three

Drinking

Water and

the Consumer

9

Bottled Water

Bottled water is everywhere. From the corporate boardrooms of New York City to the campgrounds of Yosemite National Park, Americans are drinking bottled water as never before. Once enjoyed only by the rich and privileged, bottled water is now widely available and is consumed by a large swath of the general public. In the space of fifteen years, from 1984 to the present, the average consumption of bottled water in the United States has tripled, and it shows no signs of abatement.

What accounts for this increase? What has led people to forsake cheap municipal tap water for expensive bottled water? More important, is this increased consumption justified on the grounds of health? In this chapter we explore these issues. First, however, we describe the bottled water industry.

The Bottled Water Industry

Bottled water is defined as water that is sealed in a sanitary container, is sold for human consumption, and meets all state, federal, and industry standards. It cannot contain any sweeteners or chemicals other than flavors or extracts that comprise no more than 1 percent by weight of the final product. Bottled water with more than this amount of additives is considered a soft drink and is not regulated in the same way as bottled water.

Noncarbonated bottled water is the fastest growing segment of the U.S. beverage industry. For the past ten years the growth rate of this segment has been 10 to 20 percent yearly. In 1980 the total consumption of bottled water was estimated to be 700 million gallons; in 1993 it was 2.5 billion gallons; by the year 2000 it is projected to be 4.6 billion gallons, which represents a per capita consumption of more than 15 gallons per person per year. Recent annual sales have reached 3.6 billion dollars.

Approximately seven hundred different brands of bottled water are available in the United States. Of those originating in this country, approximately 75 percent come from protected and uncontaminated wells and springs. The other 25 percent are derived from municipal water supplies that are further purified and treated prior to bottling. In addition to the waters originating in this country, there are approximately seventy-five brands that are imported. France is the country from which the most bottled water is imported. Evian and Perrier are its leading brands. Next in terms of import quantity are Canada and Italy, with such brands as Naya and San Pellegrino, respectively.

In the 1970s, prior to the widespread consumption of bottled water in this country, several of the largest bottled water companies in each region of the United States were purchased by the Perrier company. Among these companies were those selling well-known brands such as Poland Springs and Arrowhead. The Perrier company, in turn, is controlled by the Swiss conglomerate Nestle S.A.

There are several varieties of bottled water. The Food and Drug Administration requires that each bottled water carry a label identifying its variety. Among the varieties are:

Artesian Water. This variety comes from a well that taps into an aquifer, which is a water-bearing layer of rock or sand that is underground.

Mineral Water. This variety contains, at a minimum, 250 parts per million of dissolved solids. Most of these solids are minerals (such as calcium and magnesium) and trace metals.

Spring Water. This variety is derived from an underground formation from which water flows naturally to the surface. This type of water must be collected at the spring or from a borehole that taps the underground formation near where the spring derives.

Distilled Water. This water is produced by the process of distillation. Water is vaporized and then recondensed to leave it free of dissolved minerals. (Purified water also has all the solids removed, either by distillation or by other filtration processes, but it must meet certain laboratory criteria for purity. This type of water is generally used in research.)

Sparkling Water. This water is naturally carbonated due to the geothermal environment from which it derives. Certain waters are carbonated, but in processing lose their "fizz." These waters can be treated and have the carbon dioxide replaced as long as the carbon dioxide is replaced at the same concentration as that of the water when it emerged from its source.

If any of these waters is low in sodium (less than 5 milligrams per liter) they can be called sodium free.

The Regulation of the Bottled Water Industry

The Safe Drinking Water Act of 1974 placed municipal tap water under the regulation of the Environmental Protection Agency (EPA). In 1975 the Food and Drug Administration (FDA) received jurisdiction to regulate and supervise bottled water as a food product. Included in its jurisdiction was the mandate to set maximum limits on the levels of chemical, bacteriologic, and radioactive contaminants. These standards were similar to those set by the EPA for municipal water supplies. Indeed, when the EPA enacted new regulations for tap water, the FDA was required to do the same for bottled water. If the FDA did not agree with these new standards, it was required to publish the reasons in the *Federal Register.*

The FDA is also required, under the Good Manufacturing Practices laws, to oversee the processing and bottling of water. Sanitary maintenance of plants and buildings, sewage disposal, equipment and building design, and controls to ensure proper quality control all fall under its jurisdiction. It is the FDA's responsibility to ensure that there is proper labeling of the various bottled waters, such that "sparkling water" or "mineral water" are what they purport to be.

In addition to the federal government's oversight of the bottled water industry, there is state regulation. Chief among the state's responsibilities are the inspection, sampling, and analysis of approved water sources. The state also certifies testing laboratories that monitor water quality and inspects water bottling plants.

The water industry also regulates itself. As a condition of becoming a member of the International Bottled Water Association (IBWA) a bottler must agree to submit to an annual unannounced plant inspection by a recognized third-party inspection organization. Quality and compliance with state and federal regulations are also conditions of membership.

Imported waters must meet all federal and state regulations. Those imported from Europe must meet the standards set forth by the European Union as well. International bottlers that are members of the IBWA and sell water in the United States must submit inspection certificates to the IBWA.

Why Drink Bottled Water?

Marketing studies have shown that the consumer most likely to use bottled water is an adult eighteen to thirty-four years of age who is educated, upscale, and health conscious. This consumer drinks bottled water to make a statement about who he or she is and what he or she stands for. At a time when chemical pollution is ubiquitous, bottled water is perceived as untouched and pure. Therefore, bottled water has become a status symbol of health. In this regard, Californians consume approximately 40 percent of the nation's bottled water.

Although the level of purity of bottled water is high, it should be remembered that bottled water is not above contamination. Small amounts of organic chemicals and pesticides reach groundwaters. For this reason, there are low levels of organic contaminants in spring and well waters, just as there are in municipal waters that are mostly derived from surface waters. In 1990, for example, the carcinogen benzene was found in one of the leading bottled waters that is consumed in this country. Also, like tap water, bottled water is not sterile (unless ozonated) and may contain several bacteria that are usually not harmful. Because municipal tap water in this country is of such a high caliber, bottled water does not appear to offer a health advantage over tap water.

On the other hand, if minerals such as calcium and magnesium have a salutary effect on the cardiovascular system and in the prevention of osteoporosis, it is possible that bottled mineral waters may have a health advantage over municipal tap water. As described in Chapter 7 on water and minerals, the evidence to date is weak and tentative.

The one group that definitely can benefit from certain types of bottled waters are those whose immune systems are not working well, the immunocompromised. For those with AIDS or those who have had organ transplants it is important to have water that is as free as possible of parasitic infection (see Chapter 4 on drinking water and infectious diseases). This is water that has undergone reverse osmosis treatment or ozonation (see Chapter 10). For the vast majority of individuals using bottled water, such considerations are not important.

Perhaps the main advantage of bottled water over tap water is taste. This is very much an issue of individual preference, but many people find bottled water more palatable than processed municipal water. Taste in tap water is affected by the chlorine that is used to disinfect the water, which often imparts a slightly acidic taste. Certain metals, such as iron, which are derived from old pipes and storage tanks, may also contribute to taste. The taste of bottled water, on the other hand, is strongly related to its mineral content, which in turn is a product of the geological environment from which it originated. Water with a high mineral content tastes metallic, whereas that with a high bicarbonate level tastes salty. Distilled or purified water is almost always flat or dull in taste. Generally speaking, most people prefer their noncarbonated water with 30 to 100 parts per million of dissolved solids, whereas a somewhat higher concentration is preferred in carbonated water. Taste tests are held throughout the year by IBWA members to choose waters of exceptional quality.

To enhance the sense of health that bottled water conveys, most bottlers use clear plastic bottles made from polyethylene terephthalate (PET for short). Not only do these bottles look clean, but they prevent the transfer of odors through them and they keep the water's original taste and freshness. In addition, these bottles are environment friendly and can be recycled.

For further information about bottled water, one can contact the individual maker of the brand that he or she prefers (found on the bottle's label) or the International Bottled Water Association (IBWA), 113 North Henry Street, Alexandria, Virginia 22314 (telephone 800 WATER-11; fax 703 683-4074). Many companies also maintain web sites that can be consulted.

10

Water Purification

Considering all the contaminants that pollute drinking water sources and the conduits that carry water, it is not surprising that a plethora of advertisements for water filters and purifiers has appeared in the popular media in the past several years: "For clear, naturally tasting water, buy. . . ." or "For the taste of fresh mountain water, buy. . . ." In the background are pastoral scenes of majestic snow-capped mountains with running streams of (presumably) pristine waters. Health, freshness, and vitality are suggested, luring the health-conscious consumer to buy the product advertised.

In this chapter we discuss two questions related to water purification: does it make sense for the average consumer to invest in water filters or softeners, and how does one choose a filtering device?

Why Buy a Water Purifier?

Home water purifiers serve one of two purposes: either improving taste or removing unwanted contaminants. With regard to taste, drinking water from different sources may contain elements that may make it unpalatable. These elements are either organic (derived from life forms containing carbon) or inorganic (such as metals, minerals or chlorine). The sources of these contaminants depend on the water's sources or the conduits through which it runs. For example, a strong rotten egg smell suggests that source water is contaminated by organic deposits that are releasing hydrogen sulfide. An odor from hot water, but not from cold water, suggests that the problem lies in the water heater or pipes. Water with either problem could benefit from a filter, but the types of filters that would solve the two problems would differ.

With regard to removing unwanted contaminants, the type of water purifier to be used will depend on the purpose of filtration. If the desire is to purify water of infectious agents such as bacteria and parasites, certain types of mechanical filters are appropriate. On the other hand, if one desires to remove metals (e.g., lead) or minerals (e.g., calcium), resins may be needed for chemical purification of water. If excessive levels of one or more contaminants are found or one has a compromised immune system that puts him or her at risk for parasitic infections, a water treatment device should be considered.

Types of Water Purification

There are several different types of methods available for personal or home water purification.

Mechanical Filtration

Mechanical filters eliminate contaminants by forcing water through one set of or a series of small holes that do not allow substances larger than the diameter of the holes to pass through. These filters can be attached to a spigot or a carafe and catch debris that is usually more than 25 microns in diameter (a micron is one millionth of a meter; a human hair, by way of comparison, is 100 microns in diameter). Such filters can sift out debris, but are unable to remove bacteria and parasites, which are usually less than 20 microns in diameter. Moreover, they do not remove inorganic or organic chemicals, metals, or minerals. Therefore, these types of filters frequently have carbon attached to them. Carbon removes chemicals when the contaminant molecules link to them chemically (called adsorption). In such a way chlorine can be removed from the water.

The downside of such filters is that they can effectively filter only a small amount of water. They are designed so that water can go through them only at a certain flow rate, beyond which they lose their efficacy. Therefore, if one purchases such a filter one should be sure that there is a way of limiting the rate at which water passes through the spigot. In addition, with time the carbon-binding sites become filled and are no longer able to trap contaminants. Also, these filters build up deposits that are good breeding media for bacteria. They actually may end up recontaminating the very water they were meant to purify! For these reasons, mechanical filters need to be replaced regularly.

In order to filter larger quantities of water, there are cartridge-type filters. These are usually 10 to 20 inches long and can be directly attached above the counter to a faucet with a valve to divert water to them. Under-cabinet filters are directly connected to the water supply. These filters also use pores and carbon. They can filter hundreds of gallons of water before requiring a change. They can be manufactured to remove particles of varying sizes. Most remove particles down to 25 microns in diameter, but they can be made to remove particles as small as 5 microns in diameter. In such instances they may be effective in ridding water of most infectious agents, though this may still be insufficient for people with compromised immune systems. Some manufacturers add small amounts of disinfectants as well. Like all filters, these need to be replaced on a regular basis.

Ion Exchange

A second type of water purifier is based on ion exchange. These devices are most often used for the treatment of "hard" water that contains high concentrations of calcium and magnesium. However, they can be modified to remove specific substances such as nitrates, lead, or fluoride. Plastic beads, called resins, attract sodium chloride (salt), which attaches to them through chemical forces. As water passes around them the calcium, magnesium, or metal ions that are in the water dislodge the sodium ions from the beads and attach to the beads. When no sodium ions remain on the beads, they are exhausted and can no longer perform their function. Flushing the beads with saltwater restores their capacity.

This form of water treatment is relatively expensive. It is recommended only for those whose water is truly hard, which would be defined as having 150 milligrams or more of calcium per 100 milliliters of water. At this level of hardness, laundering with detergents is inefficient and pipes and appliances can become clogged with mineral deposits. Reducing the mineral content of water that has less than 150 milligrams of calcium per 100 milliliters of water is not cost effective.

When purchasing such a unit, one should be sure that it can handle the daily water needs of family members. An average individual uses about 100 gallons of water per day! This includes water for bathing, drinking, and personal hygiene. Also, one should be aware of how often a unit needs to be recharged. There are several methods for detecting the need for recharging (timers, electronic sensing devices, and computers), and one should be comfortable with the method before purchasing a unit.

In addition to cost, the other disadvantage of ion exchange resins is that they add salt to drinking water. For every atom of dissolved mineral that is removed from water by attachment to the resin, an atom of sodium is lost from the resin and added to the water. Although not much is added, this should be taken into account by people with high blood pressure, heart disease, or kidney disease. There are now several salt-free resin units.

Distillation

The third strategy for removing contaminants from water is the process of distillation. The device used is called a still. Water is boiled. The vapor is then collected and re-condensed, minus the sediment, metals, bacteria, parasites, and other contaminants that are not carried by the water vapor. Although this is an excellent method for removal of bacteria, oocysts, metals, and inorganic compounds, it does not remove all organic chemicals. Volatile organic chemicals, such as benzene, have lower boiling points than water. Therefore, they can vaporize together with steam, resulting in a distillate that is not 100 percent pure. To overcome this problem, gas vents are available in newer stills. Unfortunately, distillation is laborious and expensive. It is not economical, and only small quantities of purified water are produced. It is generally used in laboratories and industries that require high-quality water.

Reverse Osmosis

A specialized water purification process is reverse osmosis. In this process water is forced through a semipermeable membrane with extremely small pores. The membrane is generally used in combination with carbon. These filters have the capability of removing a broad range of infectious agents, inorganic and organic chemicals, and metals. Unfortunately, they have a rather limited capacity, since the movement of water through the membrane is slow. After filtration water must flow into a container, which provides a "reservoir" for the filtered water. However, because these filters are capable of removing parasitic oocysts, they are important for people with compromised immune systems. Home devices are available.

Ozonation

A form of water purification used in Europe but not often in this country is ozonation—that is, the use of ozone. Ozone is a form of oxygen in which there are three atoms (O_3) instead of the two atoms that we usually associate with oxygen (O_2). It is generated when sufficient energy comes in contact with oxygen to cause it to dissociate into two separate oxygen atoms, after which the free oxygen atoms attach to O_2 to become O_3. For example, when ultraviolet radiation from the sun hits the oxygen in the earth's upper atmosphere, ozone is formed. This process acts to absorb certain wavelengths of the sun's radiation and prevents them from reaching earth. The ozone layer thus protects us.

Using ultraviolet bulbs or electrical currents, it is possible to form ozone molecules commercially. These can then be bubbled into water. Ozone works to disinfect water by virtue of its chemical structure, which allows it to oxidize, or add oxygen, to organic compounds. In so doing, it disrupts the chemical structure of the organic compounds.

Microorganisms, which consist of organic compounds, are destroyed and water is thus sterilized.

Ozone has been shown to work much faster than chlorine. Microorganisms cannot become resistant to ozone as they can to chlorine. Besides, ozone does not combine with organic compounds found in water to form potentially dangerous byproducts (as does chlorine, which gives rise to organochlorine products). Therefore, ozonated water is very safe and can be used by people with impaired immune systems. It is not, however, widely available commercially owing to its (present) relatively high cost and the fact that, unlike chlorine, it does not maintain water sterility in the distribution system of pipes that lead from water plants. Many brands of ozonated bottled water are available.

Other devices, such as degassifiers and aerator systems, are beyond the scope of this book. They can be looked up using the world wide web at www.goodwaterco.com.

Water Purification and Camping

Camping presents a special situation with regard to water purification. Campers are at a high risk of contracting parasitic infections, since most of the surface water sources that are used by campers are also used by animals. As we mentioned in Chapter 4 on infectious diseases, many animals harbor parasites in their intestines. These parasites can cause diarrhea if ingested by humans through contaminated water.

The parasite most commonly contracted while camping is *Giardia lamblia.* Concern for such infection has led to the development of a variety of portable water filters and disinfectants for use in the wild. In one study that compared several filters head to head for the removal of *Giardia*, a great deal of inconsistency was seen (see the article by Ongerth et al in the bibliography). Nonetheless, two filters with pores of very small diameters (ultrafilters) were effective 100 percent of the time in removing oocysts, even on repeated testing. Therefore, it would behoove the camper to seek a filter with small pores.

Other methods more traditionally used to purify water while in the outdoors include the use of disinfectants. These include either iodine or chlorine-based products. Iodine disinfectants appear to be more effective than chlorine products. Still, up to eight hours of contact between the water and the iodine may be required to totally disinfect the water. Contact for thirty minutes may result in only 90 percent sterilization, which is insufficient for preventing disease. Heating water is perhaps the easiest and safest method for purifying water. Boiling water is 100 percent effective in preventing disease.

Regulation of Water Filtration

The National Sanitation Foundation (NSF), a nongovernmental not-for-profit corporation, has been given the responsibility by the federal government to establish standards for water filters. The NSF has a world wide web page (www.nsf.org) that allows consumers to look up each water filter and examine its capabilities. For a device to be approved by the NSF, certain criteria must be met: the device must demonstrate that it can reduce the contaminant for which the device was made; the materials from which the device is constructed must have been tested and found not to contribute toxic substances to the filtered water; the device must be found to be structurally sound in pressure and cycling tests; and all literature supplied with the device must be accurate. Furthermore, manufacturers of the devices must submit to periodic retesting and unannounced inspections. For an added layer of protection, the NSF also certifies faucets.

The Certified Product Database on the web page of the NSF lists certified products. From there one can contact manufacturers and distributors.

Practical Advice

No form of water treatment can yield 100 percent pure water. There will always be some form of contaminant. Purification can, however, be very effective in removing the vast majority of contaminants. The following list indicates the means that are deemed the most appropriate for removing individual contaminants or classes of contaminants. The list is not inclusive, and the web site www.goodwaterco.com can be consulted for those contaminants not mentioned here.

- Chlorine—carbon filtration.
- Nitrates—reverse osmosis.
- Fluorides—reverse osmosis.
- Iron—carbon filtration.
- Lead—carbon filtration.
- Sodium—distillation.
- Volatile organics—distillation, reverse osmosis.
- Turbidity—filtration.
- Arsenic—carbon filtration, reverse osmosis.
- Bacteria—filtration, ozonation.
- *Cryptosporidium*—ultrafiltration, ozonation, reverse osmosis.
- *Giardia*—ultrafiltration, reverse osmosis, ozonation.
- Odor—carbon filtration.
- Radon—carbon filtration.
- Organics—carbon filtration, reverse osmosis, ultrafiltration.
- Pesticides—carbon filtration, reverse osmosis.
- Radium—reverse osmosis.

Conclusion

A great deal of information has been presented in this book. At its close a few summary remarks are appropriate to synthesize the many disparate topics that we have discussed into a cohesive whole.

In the preceding chapters we have shown how the health effects of drinking water and its contaminants—and knowledge and attitudes about them—have evolved. Up until the beginning of this century, the greatest danger posed by drinking water was that of infectious diseases. The effects were immediate, and death often quickly ensued. The institution of public health measures led in large part to an abatement of these acute problems. With the longer life span that is now enjoyed as a result of the containment of infectious diseases, the focus of drinking water's potential dangers has moved from acute illness to chronic illness. Illnesses such as cancer and dementia, which take years to decades to develop, have become increasingly prevalent, and their associations with drinking water's contaminants are now becoming better appreciated. The increase in the prevalence of many of these illnesses coincided with the widespread use of chemicals after World War II, and only slowly have efforts been made to contain their pervasiveness and their consequent effects on health.

At the close of the millennium, we are now faced with the prospect that drinking water and its contaminants may have health effects that reach beyond our own lifetimes into generations to come. An effect on fertility has been suggested. Although this effect is still not clear, government agencies and academic institutions in this country and abroad are looking at this issue with concern. More time is needed to verify or discount this finding and hypothesis.

In all three phases we have seen that the contaminants that have polluted drinking water have been a reflection of society and its mores. Infected drinking waters were as much a result of ignorance as of a disregard for the health of the community and of the individual. The more recent pollution of ground and surface waters with chemicals and metals is also in part a result of ignorance, but more a consequence of hubris and disregard for the laws of nature and the environment.

Perhaps the most important lesson to be learned from all of this is that the maintenance of high-quality drinking water is a dynamic process that changes with the changes that occur in science, society, and knowledge. It is also clear that any attempt to rectify these imbalances is fraught with the danger of untoward and unknown consequences. Therefore, vigilance is and will be needed if our most vital natural resource—and consequently our continued health—is to be ensured.

Appendix 1

*Drinking Water
Contaminant Levels
Regulated by the
Federal Government*

Below is a partial list of contaminants whose levels in drinking water are regulated by federal law. Water suppliers are required to periodically check for these levels. This list is taken from the Environmental Protection Agency, Office of Water, Washington, D.C. (Form EPA 810-F-94-001, February 1994). MCLGs are maximum contaminant level goals; MCLs are maximum contaminant levels.

Contaminant	MCLG (mg/l)[1]	MCL (mg/l)	Potential Health Effect from Ingestion of Contaminated Water	Source of Water Contamination
A. Coliforms and Infectious Agents				
Total coliform	0	—	Gastroenteric disease	Human/animal fecal waste
Viruses	0	—	Gastroenteric disease	Human/animal fecal waste
Giardia lambia	0	—	Gastroenteric disease	Human/animal fecal waste

(table continues on next page)

Contaminant	MCLG (mg/l)	MCL (mg/l)	Potential Health Effect from Ingestion of Contaminated Water	Source of Water Contamination
B. Inorganic Agents				
Asbestos	7[2]	7[2]	Cancer	Asbestos cement in water pipes
Cadmium	0.005	0.005	Kidney disease	Corrosion of galvanized pipes
Chromium	0.1	0.1	Liver and kidney disease	Mining, natural deposits
Mercury	0.002	0.002	Kidney and nervous system disorders	Crop runoff, natural deposits
Nitrites	1	1	Methemoglobulinemia	Fertilizers, animal waste, natural deposits
Nitrates	10	10	Methemoglobinemia	Animal waste, septic systems, fertilizers
Fluoride	4	4	Skeletal and tooth fluorosis	Natural deposits, treatment of water
Lead	0	—	Nerve and kidney disease	Soldering, corroded plumbing
Copper	1.3	—	Gastrointestinal disease	Corroded plumbing, natural deposits
Nickel	0.1	0.1	Heart damage	Metal alloys, batteries, chemicals
Cyanide	0.2	0.2	Nerve damage	Mining, fertilizers, electroplating
C. Volatile Organic Agents				
Benzene	0	0.005	Cancer	Pesticides, paints, plastics
Carbon tetrachloride	0	0.005	Cancer	Solvents, industrial by-products
Vinyl chloride	0	0.002	Cancer	Solvents
D. Organic Agents				
Aldicarb	0.001	0	Nervous system injury	Insecticides
Carbofuran	0.04	0.04	Nervous and reproductive system injury	Soil fumigants
Chlordane	0	0.002	Cancer	Termite insecticides
Heptachlor	0	0.004	Cancer	Termite insecticides
Lindane	0.0002	0.0002	Liver, kidney, and immune system injury	Garden and lumber insecticides
PCBs	0	0.0005	Cancer	Coolant oils; plasticizers
Toluene	1	1	Liver, kidney, and nerve disease	Gasoline additives
Xylenes	10	10	Liver, kidney, and nerve disease	Gasoline by-products

[1] All units are mg/l (milligrams per liter) unless otherwise indicated.
[2] Asbestos levels are measured in units of millions of fibers per liter.

Appendix 2

Drinking Water Contaminant Levels Not under Mandatory Federal Government Regulation

Below is a list of drinking water contaminants whose levels are not under enforceable government regulation but that the federal government suggests be measured on a regular basis. States may, however, adopt their own enforceable regulations governing these contaminants. (Most measures are in mg/l, which means milligrams per liter.)

Contaminant	Suggested Level	Contaminant Effect
Aluminum	0.05–0.2 mg/l	Water discoloration
Chloride	250 mg/l	Poor taste, pipe corrosion
Copper	1 mg/l	Poor taste, porcelain staining
Fluoride	2 mg/l	Dental staining
Iron	0.3 mg/l	Poor taste, laundry staining
Manganese	0.05 mg/l	Poor taste, laundry staining
pH	6.5–8.5	Pipe corrosion
Sulfate	250 mg/l	Poor taste, laxative effect
Total dissolved solids*	500 mg/l	Poor taste, pipe corrosion; possible heart disease with low levels
Zinc	5 mg/l	Poor taste

*Water with high levels is referred to as "hard" water.

Appendix 3

Possible Cancer-Causing Contaminants of Drinking Water

Below is a list of agents that are known to cause cancer or are believed to have the potential for causing cancer. The cancer potential of these agents is determined by the International Association for Research on Cancer (IARC). See Chapter 5 for details. MCLGs are maximum contaminant level goals; MCLs are maximum contaminant levels. See Chapter 2 for a discussion of MCLGs and MCLs.

Contaminants	IARC Group	MCLG (mg/l)[1]	MCL (mg/l)	Sources of Contaminant in Drinking Water
Volatile Organic Agents				
Benzene	1	0	0.005	Some foods; gas, drugs, pesticides, paint, and plastics industries
Carbon tetrachloride	2B	0	0.005	Solvents and their degradation products
p-Dichlorobenzene	2B	0.075	0.075	Room and water deodorants and mothballs
1,2-Dichloroethane	2B	0	0.005	Leaded gasoline, fumigants, paints
1,1-Dichloroethylene	NR[2]	0.007	0.007	Plastics, dyes, perfumes, paints
Trichloroethylene	2A	0	0.005	Textiles, adhesives, metal degreasers

Contaminants	IARC Group	MCLG (mg/l)	MCL (mg/l)	Sources of Contaminant in Drinking Water
Vinyl chloride	1	0	0.002	PVC pipes, formed by solvent breakdown
Inorganic Agents				
Antimony	2B	0	0.006	Fire retardants, ceramics, electronics, fireworks, solder
Asbestos (>10 m)	1	7[3]	7[3]	Natural deposits, asbestos cement in water systems
Beryllium	1	0.004	0.004	Electrical, aerospace, and defense industries
Cadmium	1	0.005	0.005	Galvanized pipe corrosion; natural deposits, batteries, paints
Chromium	1	0.1	0.1	Natural deposits, mining, electroplating
Nitrate	NR	10	10	Animal waste, fertilizer, natural deposits, septic tanks, sewage
Nitrite	NR	1	1	Same as nitrate (rapidly converted to nitrate)
Organic Agents				
Acrylamide	2A	0	TT[4]	Polymers used in sewage/waste water treatment
Alachlor	NR	0	0.002	Herbicides on corn, soybeans, other crops
Chlordane	2B	0	0.002	Soil treatment for termites
Dibromochloro-propane	2B	0	0.002	Soil fumigants on soybeans, cotton, pineapple, orchards
Dichloromethane	2B	0	0.005	Paint strippers, metal degreasers, propellants, extraction agents
Dioxin	1	0	0.00000003	Chemical production by-products, herbicides
Epichlorohydrin	2A	0	TT	Water treatment chemicals; epoxy resins, coatings
Ethylene dibromide	2A	0	0.00005	Leaded gasoline additives; soil fumigants
Heptachlor	2B	0	0.0004	Insecticides for termites, (crops very few)
Heptachlor epoxide	NR	0	0.0002	Heptachlor
Hexchlorobenzene	2B	0	0.001	Pesticide production by-products
PAHs (benzo(a)-pyrene)	2A	0	0.0002	Coal tar coatings; burning organic matter, volcanoes, fossil fuels
PCBs	2A	0	0.0005	Coolant oils from electrical transformers; plasticizers
Phthalate (di (2-ethylhexyl))	2B	0	0.006	PVC and other plastics
Simazine	NR	0.004	0.004	Herbicides on grass sod and some crops, aquatic algae
Tetrachloroethylene	2A	0	0.005	Improper disposal of dry cleaning chemicals and other solvents

(table continues on next page)

Contaminants	IARC Group	MCLG (mg/l)	MCL (mg/l)	Sources of Contaminant in Drinking Water
Toxaphene	2B	0	0.003	Insecticides on cattle, cotton, soybeans
Other Interim Standards				
Alpha emitters	NR	0	15[5]	Decay of radionuclides in natural deposits
Arsenic	1	0.05	0.05	Natural deposits, smelters, glass, electronics wastes, orchards
Beta photon emitters	NR	0	4[6]	Radionuclides in natural and man-made deposits
Combined radium 226/228	NR	0	5[7]	Natural deposits
Total trihalo-methanes	NR	0	0.10	Drinking water chlorination by-products

[1]All units are mg/l (milligrams per liter) unless otherwise noted.
[2]NR = No rating available from IARC.
[3]Asbestos levels are measured in units of million fibers per liter.
[4]TT = Special treatment techniques required.
[5]Alpha emitter levels are measured in units of picocuries per liter.
[6]Beta photoemitter levels are measured in units of millirems (a measurement of absorbed radiation) per year.
[7]Radium levels are measured in units of picocuries per liter.

Appendix 4

Mineral Content of Commonly Used Bottled Waters

This is a partial list of the mineral and sodium contents of bottled waters commonly used in the United States and the rest of the world. Values are given in milligrams per liter. A more complete list can be obtained from *The Good Water Guide* by Maureen and Timothy Green, Rosendale Press, 1994. Alternatively, a list can be accessed from the web under www.execpc.com/~magnesum/waters.html.

Name Brand of Water	Calcium	Magnesium	Sodium
Abbey Well	54	36	45
Abode Springs	3	96	5
Alhambra	10	5	5
Apollinaris	89	104	425
Aqua Cool	45	1	3
Aqua-Pura	53	7	27
Arrowhead	20	5	3
Artesia	61	13	—

(table continues on next page)

Name Brand of Water	Calcium	Magnesium	Sodium
Belmont Springs	—	—	9
Black Mountain	25	1	8
Bru	23	23	10
Canadian Glacier	1	0	1
Canadian Spring	11	3	2
Carolina Mountain	6	—	5
Crystal Drinking	1	1	4
Crystal Geyser Sparkling Mineral	8	3	160
Crystal Rock	2	2	3
Deer Park	1	1	1
Evian	78	24	5
Georgia Mountain	2	0	—
Great Bear	1	1	3
Highland Spring	39	15	9
Kentucky Spring	80	—	19
La Croix	37	22	4
La Vie	23	8	60
Lithia Springs	120	7	680
Mendocino	310	130	240
Naya	38	20	6
Penafiel	131	41	159
Perrier	145	4	14
Poland Spring	—	2	3
Polar	13	2	9
San Benedetto	43	25	8
San Bernardo	12	1	1
Saratoga	64	7	9
Talking Rain	2	2	0
Utopia	76	17	8
Vichy Novelle	70	110	1
Vichy Springs	157	48	1095
Zephyrhills	52	7	4

Glossary

Aquifer underground geological formation that contains water.

Bacteria microorganisms of the kingdom *Prokaryotae*. They are single-celled, round, rodlike, or spiral and are characterized by having a cell wall or outer membrane.

Bioaccumulation the process by which a chemical substance becomes increasingly concentrated in the bodies of animals the further the animals are up the food chain. Low levels are found at the bottom of the food chain; higher levels are found high up on the food chain.

Bottled water water that is sealed in a sanitary container and sold for human consumption, which meets all state, federal, and industry standards. It cannot contain any sweeteners or chemicals other than flavors or extracts that can comprise no more of the final product than 1 percent by weight.

Cancer the uncontrolled growth and spread of cells. If untreated, the cells spread throughout the body and may lead to death.

Carcinogen a cancer-producing substance.

CDC   Centers for Disease Control and Prevention, the federal government agency that oversees many aspects of public health in the United States. It also does international work in public health.

Coliform count   the number of bacteria derived from human waste per cubic centimeter of water. It is a measure of the degree of water contamination.

Dementia   a general deterioration of mental capacities.

DES   diethylstilbestrol, a synthetic chemical with estrogen-like effects that has been shown to have deleterious effects on a fetus exposed to it during gestation.

Distillation   the process of boiling or vaporizing water and then recondensing it. This is one method for ridding water of contaminants.

DNA   deoxyribonucleic acid, the substance that makes up the chemical building blocks inside the nucleus of a cell that direct cell function and how the cell divides.

EPA   Environmental Protection Agency, the federal government agency that oversees the protection of the environment, including air and water quality and the disposal of toxic materials.

Estrogen   a hormone produced in the ovaries of women that is released into the blood and exerts its effects on various organs of the body. Estrogen is needed for the normal maturation of a female, stimulating growth of the uterus and vagina as well as secondary sex characteristics, such as breast development.

Estrogenlike molecule   a molecule that does not bear a chemical likeness to the estrogen molecule, but can nonetheless interact with the receptor for estrogen inside the cell and bring about an estrogen effect.

FDA   Food and Drug Administration, the federal government agency that, among its many functions, oversees the bottled water industry.

Fetus an unborn human that grows in the uterus after it has attached itself to the uterus wall. Generally called a fetus (as opposed to an embryo) after the eighth week of conception.

Fluoridation the process whereby fluoride is intentionally added to drinking water. This is done to protect against tooth decay.

Gestation the period of development of a fetus in the uterus during pregnancy.

Groundwater water derived from underground sources, such as aquifers, into which wells have been drilled or springs from which water naturally comes to the surface from underground.

Hardness a term used to denote the concentration of dissolved solids in water. Most often the term refers to the concentration of calcium, magnesium, and trace metals in water.

Hormone a molecule synthesized and excreted by specialized cells in the body that are released into the bloodstream and exert their biochemical effects on target cells at a distance from the site of their production and excretion.

IARC International Association of Research on Cancer. Among its functions it assesses the cancer-promoting effect of various chemicals.

IBWA International Bottled Water Association, the umbrella group for the bottled water industry.

Immune system the integrated system of white blood cells and antibodies that protects the body from infection.

Infectious disease an illness brought on by invasion of the body by bacteria, parasites, or viruses.

Inorganic molecules molecules not containing carbon atoms. In the context of this book, it refers to minerals, metals, and non–carbon-containing chemicals.

Liter the equivalent of 1.0567 quarts. A measure of volume used in the metric system.

Microbes microscopic organisms. A general term used to describe bacteria, parasites, or viruses.

Microgram one millionth of a gram.

Micron one millionth of a meter. A meter is 39.37 inches.

Milligram one thousandth of a gram. There are 454 grams in a pound.

Milliliter one thousandth of a liter.

Mineral water water that contains, at a minimum, 250 parts per million of dissolved solids. Most of these solids are minerals (such as calcium and magnesium) and trace metals.

Nervous system the integrated network of nerves found in the brain, the spinal cord, and the periphery of the body that controls the workings of muscles and internal organs.

NSF the National Sanitation Foundation (NSF), a non-governmental, not-for-profit corporation that establishes standards for water filters.

Oocyst the encapsulated form of a phase in the development of a parasite. The encapsulation often allows the developing parasite to withstand disinfection.

Organic molecules molecules that contain carbon atoms. Often they are derived from decomposed vegetation and from pollution of human origin.

Organochlorine a carbon-containing compound to which chlorine molecules are attached. Such compounds occur when chlorine used to disinfect water attaches to carbon atoms from organic compounds found in water.

Ozonation a process for purifying water whereby the gas ozone is passed through the water.

PCBs polychlorinated biphenyls, compounds found in hydraulic fluids, adhesives, and flame retardants. These chemicals have been implicated in cancer promotion and decreased fertility.

Parasites one-celled members of the kingdom *Protista*. They are larger than bacteria, but a microscope is still needed to see them. They exist ubiquitously in nature and in water and can infect humans.

Pathogens disease-producing microbes. Generally this term is used as another name for bacteria, viruses, or parasites that cause illness.

RNA ribonucleic acid. A nucleic acid that controls protein synthesis in all living cells and takes the place of DNA in certain viruses. It differs from DNA by the type of sugar that is attached to it and by having a different type of nucleic acid present.

Radionuclides elements found in nature that emit energy. They include uranium, radon, and radium and have been implicated in causing cancer.

Receptor a protein on the surface of a cell to which a chemical, such as a hormone, may attach. Once attached to the receptor, the cell receives a signal to undergo some sort of change.

Reverse osmosis a process of water purification whereby water is forced through a semipermeable membrane with extremely small pores. It has the capability of removing a broad range of infectious agents, inorganic and organic chemicals, and metals. It has a limited capacity, since the movement of water through the membrane is slow.

SDWA Safe Drinking Water Act, an act of Congress enacted in 1974 to protect water sources and drinking water. It set up government oversight of surface and groundwater sources, programs for the development of standards and regulations, and funding for state water systems. Ongoing monitoring to ensure compliance was made an integral part of these programs. There have been amendments to this legislation in the years since its enactment.

Spring water water derived from underground formations from which the water flows naturally to the surface. This type of water may be collected at the spring or from a borehole that taps the underground formation near where the spring derives.

Surface water   water derived from lakes, reservoirs, and rivers.

THM   trihalomethane, an organic compound to which inorganic atoms, such as chlorine, are attached. They are formed in water when organic compounds come in contact with chlorine that is used to disinfect water.

Virus   the smallest known infectious organism. It consists of a coat of protein enclosing a central core of nucleic acids called DNA or RNA. Through attachment to and control of its host's DNA it is able to replicate.

Bibliography

This bibliography is not inclusive or exhaustive. Rather it lists sources of information and publications that will give the reader adequate references so he or she can, if desired, further research the topics covered in this book. Owing to the nature of this book, many of the references are to medical textbooks and journals, and these can most easily be found in a medical library.

Sources of General Information

The American Water Works Association (AWWA) is an umbrella group for the water works industry. It is a nonprofit scientific and educational society dedicated to the improvement of drinking water quality and supply. It includes 50,000 members, ranging from water plant operators and managers to academicians to environmentalists and scientists. The AWWA maintains an address on the world wide web (from now on called the Web or www) from which one can obtain a great deal of up-to-date information regarding research, legislation, and newer technologies. Addresses are www.awwa.org, or AWWA, 6666 West Quincy Avenue, Denver, CO 80235 (telephone: 800 926-7337).

The Environmental Protection Agency (EPA), the government agency charged with regulating water safety, also maintains a web address from which information on current government initiatives, public hearings and forums, and health alerts can be obtained. There are many subdirectories dealing with each subject individually. Addresses: www.epa.gov, or EPA, Office of Groundwater and Drinking Water, U.S. EPA (4601), 401 M Street, S. W., Washington, D.C. 20460 (telephone: 800 426-4791).

Information regarding the toxic effects of various chemicals can be obtained from the Toxicology Data Network (TOXNET), which is a computerized system of files oriented to toxicology. One subfile, the Toxic Release Inventory, is specific as to how much toxic material is released by industry into the environment. Toxicology Information On-Line (TOXLINE/TOXLIT) is available from the National Library of Medicine in Washington, D.C., and offers comprehensive information on toxicology. Address: www.nlm.nih.gov (telephone: 301 496-6308).

Those who want more medical information can access MEDLINE from the National Library of Medicine in Washington, D.C. Specific medical subjects can be referenced from more than three thousand medical journals. Address: www.nlm.nih.gov (telephone: 301 496-6308).

For more information on bottled water, one can contact the International Bottled Water Association (IBWA). Addresses are www.bottledwater.org, or IBWA, 1700 Diagonal Road, Suite 650, Alexandria, VA 22314 (telephone: 800 928-3711).

Books of General Information

Long Cohuna Kupua A'o, L. *Don't Drink the Water: The Essential Guide to Our Contaminated Drinking Water and What You Can Do About It.* Kali Press, Pagosa Springs, Colorado, 1997.

National Academy of Sciences. *Drinking Water and Health.* National Academy Press, Washington, D.C., 1977.

Symons, J. *Drinking Water: Refreshing Answers to All Your Questions.* Texas A & M University Press, College Station, Texas, 1995.

1 A Brief History of Drinking Water, Sanitation, and Disease

Several books on medical history that contain references to water are:

Garrison, F. H. *An Introduction to the History of Medicine,* 4th edition. W. B. Saunders Company, Philadelphia and London, 4th edition, 1929.

Major, R. H. *A History of Medicine.* Charles C. Thomas Publisher, Springfield, Illinois, 1954.

Siegrist, H. E. *Landmarks in the History of Hygiene.* Oxford University Press, London, 1956.

Wain, H. *A History of Preventive Medicine.* Charles C. Thomas Publisher, Springfield, Illinois, 1970.

Biblical and Classical Times

Johnstone, H. W. *Private Life of the Romans.* Scott Publishers, Chicago, 1903.

Kee, H. C. *Medicine, Miracle, and Magic in New Testament Times.* Cambridge University Press, Cambridge and New York, 1986.

Kottek, S. S. Gems from the Talmud: Public Health I—Water Supply. *Israel Journal of Medical Sciences,* volume 31, pages 255–256, 1995.

Lloyd, G. E. The Fielding Garrison Lecture: The transformation of ancient medicine. *Bulletin of the History of Medicine,* volume 66, pages 114–132, 1992.

Preuss, J. *Biblical and Talmudic Medicine.* Translated and edited by Fred Rosner. Sanhedrin Press, New York, 1978.

Smith, W. D. Notes on ancient medical historiography. *Bulletin of the History of Medicine,* volume 63, pages 73–109, 1989.

The Middle Ages through the Eighteenth Century

Amundsen, D. W. Medicine and faith in early Christianity. *Bulletin of the History of Medicine,* volume 56, pages 326–350, 1982.

Bruce-Chwatt, L. J. A medieval glorification of disease and death. *Medical History,* volume 16, pages 76–77, 1972.

Carmichael, A. G. Plague legislation in the Italian Renaissance. *Bulletin of the History of Medicine*, volume 57, pages 508–525, 1983.

Ferngren, G. B. Early Christianity as a religion of healing. *Bulletin of the History of Medicine*, volume 66, pages 1–15, 1992.

King, L. S. Medical theory and practice at the beginning of the eighteenth century. *Bulletin of the History of Medicine*, volume 46, pages 1–15, 1972.

Murphy, T. D. The French medical profession's perception of its social function between 1776 and 1830. *Medical History*, volume 23, pages 259–278, 1979.

Risse, G. B. Epidemics and medicine: the influence of disease on medical thought and practice. *Bulletin of the History of Medicine*, volume 53, pages 505–519, 1979.

Sterns, I. Care of the sick brothers by the crusader order in the Holy Land. *Bulletin of the History of Medicine*, volume 57, pages 43–69, 1983.

Tourney, G. The physician and witchcraft in restoration England. *Medical History*, volume 16, pages 143–155, 1972.

Woodings, A. F. The medical resources and practice of the crusader states in Syria and Palestine 1096–1193. *Medical History*, volume 15, pages 268–277, 1971.

The Nineteenth Century

Amulree, L. Hygienic conditions in ancient Rome and modern London. *Medical History*, volume 17, pages 244–255, 1973.

DeKrief, P. *Microbe Hunters*. Harcourt, New York, 1926.

Hamlin, C. Edward Frankland's early career as London's official water analyst, 1865–1876: The context of "previous sewage contamination." *Bulletin of the History of Medicine*, volume 56, pages 57–76, 1982.

Hardy, A. Cholera, quarantine, and the English preventive system, 1850–1895. *Medical History*, volume 37, pages 250–269, 1993.

Hardy, A. Water and the search for public health in London in the eighteenth and nineteenth centuries. *Medical History*, volume 28, pages 250–282, 1984.

Lipschutz, D. E. The water question in London. *Bulletin of the History of Medicine*, volume 42, pages 510–526, 1968.

Luckin, W. The final catastrophe: Cholera in London, 1866. *Medical History*, volume 21, pages 32–42, 1977.

2   The Modern Era of Drinking Water Regulation

History

Bilson, G. The first epidemic of Asiatic cholera in lower Canada, 1832. *Medical History*, volume 21, pages 411–433, 1977.

Brewer, P. W. Voluntarism on trial: St Louis' response to the cholera epidemic of 1849. *Bulletin of the History of Medicine*, volume 49, pages 102–122, 1975.

Ellis, J. H. Businessmen and public health in the urban South during the nineteenth century: New Orleans, Memphis, and Atlanta. *Bulletin of the History of Medicine,* volume 44, pages 197–212, 346–71, 1970.

Marcus, A. I. Disease prevention in America: From a local to a national outlook, 1880–1910. *Bulletin of the History of Medicine,* volume 53, pages 184–203, 1979.

Powell, J. H. *Bring Out Your Dead: The Great Plague of Yellow Fever in Philadelphia in 1793.* University of Pennsylvania Press, Philadelphia, 1949.

Rosenberg, C. E. *The Cholera Years: The United States in 1832, 1849, and 1866.* University of Chicago Press, Chicago, Illinois, 1962.

Williams, R. C. *The United States Public Health Service, 1798–1950.* Government Printing Office, Washington, D.C., 1951.

Regulation

Auerbach, J. Costs and benefits of current SDWA regulations. *Journal of the American Water Works Association,* volume 86, pages 69–78, 1994.

National Academy of Sciences. *Drinking Water and Health.* National Academy Press, Washington, D.C., 1977. [Summarizes many aspects of the Safe Drinking Water Act.]

Raucher, R. S., Drago, J. A., Castillo, E. T., et al. *Estimating the Cost of Compliance with Drinking Water Standards: A User's Guide for Developing or Auditing Estimates of Water Supply Regulatory Compliance Costs. Final Report.* American Water Works Association Research Foundation, Denver, Colorado, 1996.

Raucher, R. S. Public health and regulatory considerations of the Safe Drinking Water Act. *Annual Review of Public Health,* volume 17, pages 179–202, 1996.

3 Drinking Water Sources, Treatment, Safety, and Conservation

Sources of information regarding water safety, processing, and distribution include:

American Water Works Association, Public Affairs Department, 6666 West Quincy Avenue, Denver, CO 80235 (telephone: 303 347-6284; web site: www.awwa.org).

Association of Metropolitan Water Agencies, 1717 K Street, N.W., Suite 1102, Washington, D.C. 20036 (telephone: 202 331-2820).

Association of State Drinking Water Administrators, 1120 Connecticut Avenue, N.W., Washington, D.C. 20036 (telephone: 202 293-7655; web site: www.asdwa.org).

National Association of Water Companies, 1725 K Street, N.W., Suite 1212, Washington, D.C. 20006 (telephone: 202 833-8383).

National Drinking Water Clearinghouse, West Virginia University, P.O. 6064, Morgantown, WV 26506–6064 (telephone: 800 624-8301; web site: www.ndwc.wvu. edu).

U.S. EPA, 401 M Street, S.W., Washington, D.C. 20460 (telephone: 202 260-7786; Safe Drinking Water Act Hotline 800 426-4791; web site: www.epa.gov).

U.S. Geological Survey, Hydrologic Information Unit, 419 National Center, Reston, VA 22092 (telephone: 703 648-6818; web site: www.water.usgs. gov).

Other sources of information on the water industry and water use are:

Anonymous. For second time in 3 years, water rationed in San Juan. *New York Times,* June 22, 1997.

Passell, P. A gush of profits from water sale? *New York Times,* April 23, 1998.

Pollack, A. Tightening the faucet: How U.S. Filter is consolidating the waterworks. *New York Times,* April 23, 1998.

Stevens, W. K. Expectations aside, water use in the U.S. is showing decline. *New York Times,* November 10, 1998.

A summary of drinking water regulation and enforcement can be found in the following article, which is available on the web page of the American Water Works organization (www.awwa.org) or on the web page of *USA Today* (www.usatoday.com):

Eisler, P., Hansen, B., Davis, A. Lax oversight raises tap water risks. *USA Today,* October 21, 1998.

4   *Drinking Water and Infectious Diseases*

General information on infectious agents is available on the web site of the American Society of Microbiology, www.amusa.com.

Cholera

Besser, R. E., Feikin, D. R., Eberhart-Phillips, J. E., et al. Diagnosis and treatment of cholera in the United States: Are we prepared? *Journal of the American Medical Association,* volume 272, pages 1203–1205, 1994.

Centers for Disease Control and Prevention. Update: Vibrio cholera 01—Western hemisphere, 1991–1994, and V. cholerae 0139—Asia, 1994. *Morbidity and Mortality Weekly Reports,* volume 44, pages 215–217, 1995.

Centers for Disease Control. Update: Vibrio cholerae 01—Western Hemisphere, 1992. *Morbidity and Mortality Weekly Reports,* volume 42, pages 89–91, 1993.

Marston, W. In Peru's shantytowns, cholera comes by the bucket. *New York Times,* December 8, 1998.

Wilson, M., and Chelala, C. Cholera is walking south. *Journal of the American Medical Association,* volume 272, pages 1226–1227, 1994.

E. Coli

Anonymous. The killer germ: *E. Coli. Time,* August 3, 1998.

Besser, R. E., Lett, S. M., Weber, J. T., et al. An outbreak of diarrhea and hemolytic uremic syndrome from *Escherichia coli* 0157:H7 in fresh-pressed apple cider. *Journal of the American Medical Association,* volume 269, pages 2217–2220, 1993.

Feng, P. *Escherichia coli* serotype 0157:H7: Novel vehicles of infection and emergence of phenotypic variants. www.cdc.gov/ncidod/ EID/vol1no2.

Griffin, P. M., and Tauxe, R. V. The epidemiology of infections caused by *Escherichia coli* 0157:H7, other enterohemorrhagic *E. coli,* and the associated hemolytic uremic syndrome. *Epidemiology Reviews,* volume 13, pages 60–98, 1991.

Keene, W. E., McAnulty, J. M., Hoesly, F. C., et al. A swimming associated outbreak of hemorrhagic colitis caused by *Escherichia coli* 0157:H7 and *Shigella sonnei. New England Journal of Medicine,* volume 331, pages 579–584, 1994.

Swerdlow, D. L., Woodruff, B. A., Brady, R. C., et al. A waterborne outbreak in Missouri of *Escherichia coli* 0157:H7 associated with bloody diarrhea and death. *Annals of Internal Medicine,* volume 117, pages 812–819, 1992.

Campylobacter

Blaser, M. J., Berkowitz, I. D., Laforce, F. M., et al. *Campylobacter* enteritis in the United States: A multicenter study. *Annals of Internal Medicine,* volume 98, pages 360–365, 1983.

Vogt, R. L., Sours, H. E., Barrett, T., et al. *Campylobacter* enteritis associated with contaminated water. *Annals of Internal Medicine,* volume 96, pages 292–296, 1982.

Salmonella

Kramer, M. H., Herwaldt, B. L., Craun, G. F., et al. Surveillance of waterborne-disease outbreaks—United States, 1993–1994. *Morbidity and Mortality Weekly Reports,* volume 45, SS-1, pages 1–45, 1996.

Shigella

Kramer, M. H., Herwaldt, B. L., Craun, G. F., et al. Surveillance of waterborne-disease outbreaks—United States, 1993–1994. *Mor-*

bidity and Mortality Weekly Reports, volume 45, SS-1, pages 1–45, 1996.

Levine, W. C., Stephenson, W. T., and Craun, G. F. Waterborne disease outbreaks, 1986–1988. *Morbidity and Mortality Weekly Reports,* volume 39, SS-1, pages 1–9, 1990.

St. Louis, M. E. Water-related disease outbreaks, 1985. *Morbidity and Mortality Weekly Reports,* volume 37, SS-2, pages 15–24, 1988.

Cryptosporidium

Anonymous. Drinking water priority rulemaking: Microbial and disinfection byproduct rules. www.epa.gov/OGWDW/mdbp/mdbp.

DuPont, H. L., Chappell, C. L., Sterling, C. R., et al. The infectivity of *Cryptosporidium parvum* in healthy volunteers. *New England Journal of Medicine,* volume 332, pages 855–859, 1995.

Fayer, R. Effect of high temperature on infectivity of *Cryptosporidium parvum* oocysts in water. *Applied Environmental Microbiology,* volume 60, pages 2732–2735, 1994.

Gallaher, M. M., Herndon, J. L., Nims, L. J., et al. Cryptosporidiosis and surface water. *American Journal of Public Health,* volume 79, pages 39–42, 1989.

Goldstein, S. T., Juranek, D. D., Ravenholt, O., et al. Cryptosporidiosis: An outbreak associated with drinking water despite state-of-the-art water treatment. *Annals of Internal Medicine,* volume 124, pages 459–468, 1996.

Guerrant, R. L. Cryptosporidiosis: An emerging, highly infectious threat. *Emerging Infectious Diseases,* volume 3, pages 51–57, 1997.

Hayes, E. B., Matte, T. D., O'Brien, T. R., et al. Large community outbreak of cryptosporidiosis due to contamination of a filtered public water supply. *New England Journal of Medicine,* volume 320, pages 1372–1376, 1989.

Juranek, D. D. Cryptosporidiosis: Sources of infection and guidelines for prevention. www.cdc.gov/ncidod/diseases/crypto/sources.

Millard, P. S., Gensheimer, K. F., Addiss, D. G., et al. An outbreak of cryptosporidiosis from fresh-pressed apple cider. *Journal of the American Medical Association,* volume 272, pages 592–596, 1994.

Rose, J. B., Garbo, C. P., and Jakubowski, W. Survey of potable water supplies for *Cryptosporidium* and *Giardia. Environmental Sciences and Technology,* volume 25, pages 1393–1400, 1991.

Shenon, P. Proud Sydney is aghast: Its tapwater is unsafe. *New York Times,* August 2, 1998.

Sorvillo, F. J., Fujioka, K., Nahlen, B., et al. Swimming—associated cryptosporidiosis. *American Journal of Public Health,* volume 82, pages 742–744, 1992.

Giardia

Babb, R. R. Giardiasis: Taming this pervasive parasitic infection. *Postgraduate Medicine,* volume 98, pages 155–158, 1995.

Craun, G. T. Waterborne giardiasis in the United States: A review. *American Journal of Public Health,* volume 69, pages 817–819, 1979.

Marshall, M. M., Naumovitz, D., Ortega, Y., et al. Waterborne protozoal pathogens. *Clinical Microbiology Review,* volume 10, pages 67–85, 1997.

Steiner, T. S., Thielman, N. M., and Guerrant, R. L. Protozoal agents: What are their dangers for the public water supply? *Annual Review of Medicine,* volume 48, pages 329–340, 1997.

Cyclospora

Centers for Disease Control. Update: Outbreaks of cyclosporiasis—United States, 1997. *Morbidity and Mortality Weekly Reports,* volume 46, pages 461–462, 1997.

Connor, B. A., Shlim, D. R. Food borne transmission of *Cyclospora. Lancet,* volume 346, page 1634, 1995.

Herwaldt, B. L., and Ackers, M.-L. An outbreak in 1996 of cyclosporiasis associated with imported raspberries. *New England Journal of Medicine,* volume 336, pages 1548–1556, 1997.

Obi, W. E., Zimmerman, S. K., and Needham, C. A. *Cyclospora* species as a gastrointestinal pathogen in immunocompetent hosts. *Journal of Clinical Microbiology,* volume 33, pages 1267–1269, 1995.

Viruses

Ho, M., Glass, R. I., Pinsky, P. F., et al. Diarrheal deaths in American children: Are they preventable? *Journal of the American Medical Association,* volume 260, pages 3281–3285, 1988.

Hopkins, R. S., Gaspard, G. B., Williams, F. P. J., et al. A community waterborne gastroenteritis outbreak: Evidence for rotavirus as the agent. *American Journal of Public Health,* volume 74, pages 263–265, 1984.

Kramer, M. H., Herwaldt, B. L., Craun, G. F., et al. Surveillance of waterborne-disease outbreaks—United States, 1993–1994. *Morbidity and Mortality Weekly Reports,* volume 45, SS-1, pages 1–45, 1996.

LeBaron, C. W., Furutan, N. P., Lew, J. F., et al. Viral agents of gastroenteritis: public health importance and outbreak management. *Morbidity and Mortality Weekly Reports,* volume 39, RR-5, pages 1–24, 1990.

Wenman, W. M., Hinde, D., Feltham, S., et al. Rotavirus infection in adults: Results of a prospective family study. *New England Journal of Medicine,* volume 301, pages 303–306, 1979.

Pfiesteria

Glasgow, H. B. Jr., Burkholder, J. M., Schmechel, D. E., et al. Insidious effects of a toxic estuarine dinoflagellate on fish survival and human health. *Journal of Toxicology and Environmental Health,* volume 46, pages 510–522, 1995.

Grattan, L. M., Oldach, D., Perl, T. M., et al. Learning and memory difficulties after environmental exposure to waterways containing toxin-producing *Pfiesteria* or *Pfiesteria*-like dinoflagellates. *Lancet,* volume 352, pages 532–539, 1998.

Martin, J. P. *Pfiesteria:* A mangy, mysterious microbe. *Washington Post,* March 5, 1998.

Pfiesteria piscicida. www.epa.gov/OWOW/estuaries/pfiesteria.

University of Maryland. About *Pfiesteria piscicida.* www.mdsg.edu/fish-health/pfiesteria.

Warrick, J., and Brown, D. One scary, mysterious microbe. *Washington Post,* September 18, 1997.

5   Drinking Water and the Risk of Cancer

General Reviews

Black, J. J., and Baumann, P. C. Carcinogens and cancers in freshwater fishes. *Environmental Health Perspectives,* volume 90, pages 27–33, 1991.

Cantor, K. P. Drinking water and cancer. *Cancer Causes and Control,* volume 8, pages 291–308, 1997.

Clark, R. M., Goodrich, J. A., and Deininger, R. A. Drinking water and cancer mortality. *Science of the Total Environment,* volume 53, pages 153–172, 1986.

Cotruvo, J. A. Organic micropollutants in drinking water: An overview. *Science of the Total Environment,* volume 47, pages 7–26, 1985.

Dayan, A. D. Carcinogenicity and drinking water. *Pharmacology and Toxicology,* volume 72, supplement 1, pages 108–115, 1993.

Griffith, J., Duncan, R. C., Riggan, W. B., et al. Cancer mortality in U.S. counties with hazardous waste sites and ground water pollution. *Archives of Environmental Health,* volume 44, pages 69–74, 1989.

Koivusalo, M., Vartiainen, T., Hakulinen, T. et al. Drinking water mutagenicity and leukemia, lymphomas, and cancers of the liver, pancreas, and soft tissue. *Archives of Environmental Health,* volume 50, pages 269–276, 1995.

Morin, M. M., Sharrett, A. R., Bailey, K. R., et al. Drinking water source and mortality in U.S. cities. *International Journal of Epidemiology,* volume 14, pages 254–264, 1985.

Morris, R. D. Drinking water and cancer. *Environmental Health Perspectives*, volume 103, supplement 8, pages 225–232, 1995.

Ozonoff, D., Longnecker, M. P. Epidemiologic approaches to assessing human cancer risk from consuming aquatic food resources from chemically contaminated water. *Environmental Health Perspectives*, volume 90, pages 141–146, 1991.

Peeters, E. G. The influence of soil components and drinking water on the appearance of cancer: A review. *Journal of Environmental Pathology, Toxicology and Oncology*, volume 11, pages 201–204, 1992.

Shy, C. M. Chemical contamination of water supplies. *Environmental Health Perspectives*, volume 62, pages 399–406, 1985.

Arsenic

Anonymous. Chemically contaminated aquatic food resources and human cancer risk: Retrospective. *Environmental Health Perspectives*, volume 90, pages 149–154, 1991.

Bates, M. N., Smith, A. H., and Hopenhayn-Rich, C. Arsenic ingestion and internal cancers: A review. *American Journal of Epidemiology*, volume 135, pages 462–476, 1992.

Bearak, B. New Bangladesh disaster: Wells that pump poison. *New York Times*, November 10, 1998.

Bearak, B. Sounding the alarm on deadly wells. *New York Times*, December 8, 1998.

Cantor, K. P. Arsenic in drinking water: How much is too much? *Epidemiology*, volume 7, pages 113–115, 1996.

Chappell, W. R., Beck, B. D., Brown, K. G., et al. Inorganic arsenic: A need and an opportunity to improve risk assessment. *Environmental Health Perspectives* 1997, volume 105, pages 1060–1067, 1997.

Chiou, H. Y., Hsueh, Y. M., Liaw, K. F., et al. Incidence of internal cancers and ingested inorganic arsenic: A seven year follow-up study in Taiwan. *Cancer Research*, volume 55, pages 1296–1300, 1995.

Smith, A. H., Hopenhayn-Rich, C., Bates, M. N., et al. Cancer risks from arsenic in drinking water. *Environmental Health Perspectives*, volume 97, pages 259–267, 1992.

Stohrer, G. Arsenic: Opportunity for risk assessment. *Archives of Toxicology*, volume 65, pages 525–531, 1991.

Asbestos

Anonymous. Report on cancer risks associated with the ingestion of asbestos. *Environmental Health Perspectives*, volume 72, pages 253–265, 1987.

Howe, H. L., Wolfgang, P. E., Burnett, W. S. Cancer incidence following exposure to drinking water with asbestos leachate. *Public Health Reports*, volume 104, pages 251–256, 1989.

Kanarek, M. S. Epidemiological studies on ingested mineral fibres: Gastric and other cancers. *IARC Scientific Publications*, volume 90, pages 428–437, 1989.

MacRae, J. D. Asbestos in drinking water and cancer. *Journal of the Royal College of Physicians of London*, volume 22, pages 7–10, 1988.

Radionuclides

Collman, G. W., Loomis, D. P., and Sandler, D. P. Childhood cancer mortality and radon concentration in drinking water in North Carolina. *British Journal of Cancer*, volume 63, pages 626–629, 1991.

Cothern, C. R., Lappenbusch, W. I., and Michel, J. Drinking-water contribution to natural background radiation. *Health Physics*, volume 50, pages 33–47, 1986.

Cross, F. T., Harley, N. H., and Hofmann, W. Health effects and risks from 222Rn in drinking water. *Health Physics*, volume 48, pages 649–670, 1985.

Finkelstein, M. M., and Kreiger, N. Radium in drinking water and the risk of bone cancer among Ontario youths: A second study and combined analysis. *Occupational and Environmental Medicine*, volume 53, pages 305–311, 1996.

Fuortes, L., McNutt, L. A., and Lynch, C. Leukemia incidence and radioactivity in drinking water in 59 Iowa towns. *American Journal of Public Health*, volume 80, pages 1261–1262, 1990.

Lyman, G. H., Lyman, C. G., and Johnson, W. Association of leukemia with radium underwater contamination. *Journal of the American Medical Association*, volume 254, pages 621–626, 1985.

Nitrates

Beresford, S. A. A. Is nitrate in the drinking water associated with the risk of cancer in the urban UK? *International Journal of Epidemiology*, volume 14, pages 57–63, 1985.

Chilvers, C., Inskip, H., Caygill, C., et al. A survey of dietary nitrate in well-water users. *International Journal of Epidemiology*, volume 13, pages 324–331, 1984.

Geleperin, M. D., Moses, V. J., and Fox, G. Nitrate in water supplies and cancer. *Illinois Medical Journal*, volume 149, pages 251–253, 1976.

Johnson, C. J., and Kross, B. C. Continuing importance of nitrate contamination of groundwater and wells in rural areas. *American Journal of Industrial Medicine*, volume 18, pages 449–456, 1990.

Juhasz, L., Hill, M. J., and Nagy, G. Possible relationship between nitrate in drinking water and incidence of stomach cancer, *IARC Scientific Publications*, number 31, pages 619–623, 1980.

Morales Suarez-Varela, M. M., Llopis-Gonzalez, A., Tejerizo-Perez, L. Impact of nitrates in drinking water on cancer mortality in Valencia, Spain. *European Journal of Epidemiology*, volume 11, pages 15–21, 1995.

Mueller, D. K., Hamilton, P. A., Helsel, D. R., et al. Nutrients in groundwater and surface water of the United States: An analysis of data through 1992. In *Water-Resources Investigations Report 95–4031.* U.S. Geological Survey, Denver, Colorado, 1995.

Rademacher, J. J., Young, T. B., and Kanarek, M. S. Gastric cancer mortality and nitrate levels in Wisconsin drinking water. *Archives of Environmental Health,* volume 47, pages 292–297, 1992.

Ward, M. H., Mark, S. D., Cantor, K. P., et al. Drinking water nitrate and risk of non-Hodgkin's lymphoma. *Epidemiology,* volume 7, pages 465–471, 1996.

Organic Chemicals

Budnick, I. D., Sokal, D. C., Falk, H., et al. Cancer and birth defects near the Drake superfund site, Pennsylvania. *Archives of Environmental Health,* volume 39, pages 409–413, 1984.

Cohn, P., Klotz, J., Bove, F., et al. Drinking water contamination and the incidence of leukemia and non-Hodgkin's lymphoma. *Environmental Health Perspectives,* volume 102, pages 556–561, 1994.

Cutler, J. J., Parker, G. S., Rosen, S., et al. Childhood leukemia in Woburn, Massachusetts. *Public Health Reports,* volume 101, pages 201–205, 1986.

Griffith, J., Duncan, R. C., Riggan, W. B., et al. Cancer mortality in U.S. counties with hazardous waste sites and ground water pollution. *Archives of Environmental Health,* volume 44, pages 69–74, 1989.

Heath, C. W., Nadel, M. R., Zack, M. M., et al. Cytogenetic findings in persons living near the Love Canal. *Journal of the American Medical Association,* volume 251, pages 1437–1440, 1984.

Janerich, D. T., Burnett, W. S., Feck G, et al. Cancer incidence in the Love Canal area. *Science,* volume 212, pages 1404–1407, 1981.

Lagakos, S. W., Wessen, B. J., and Zelen, M. An analysis of contaminated well water and health effects in Woburn, Massachusetts. *Journal of the American Statistical Association,* volume 81, pages 583–596, 1986.

Najem, G. R., Louria, D. B., Lavenhar, M. A., et al. Clusters of cancer mortality in New Jersey municipalities; with special reference to chemical toxic waste disposal sites and per capita income. *International Journal of Epidemiology,* volume 14, pages 528–537, 1985.

Ritter, W. F. Pesticide contamination of ground water in the United States: A review. *Journal of Environmental Science and Health,* volume 25, pages 1–29, 1990.

U.S. Environmental Protection Agency. *National Survey of Pesticides in Drinking Water Wells: Phase I Report.* EPA, Washington, D.C., 1990.

Wong, O., Morgan, R. W., Whorton, M. D., et al. Ecological analyses and case control studies of gastric cancer and leukemia in relation to

DBCP in drinking water in Fresno County, California. *British Journal of Industrial Medicine*, volume 46, pages 521–528, 1989.

Chlorination Byproducts

Bull, R. J., Birnbaum, L. S., Cantor, K. P., et al. Symposium overview: Water chlorination: Essential process or cancer hazard? *Fundamental and Applied Toxicology*, volume 28, pages 155–166, 1995.

Cantor, K. P., Lynch, C. F., Hildesheim, M. E., et al. Drinking water source and chlorination byproducts. I. Risk of bladder cancer. *Epidemiology*, volume 9, pages 21–28, 1998.

Cap, A. P. The chlorine controversy. *International Archives of Occupational Environmental Health*, volume 68, pages 455–458, 1996.

Doyle, T. J., Zheng, W., Cerhan, J. R., et al. The association of drinking water source and chlorination by-products with cancer incidence among postmenopausal women in Iowa: A prospective cohort study. *American Journal of Public Health*, volume 87, pages 1168–1176, 1997.

Hildesheim, M. E., Cantor, K. P., Lynch, C. F., et al. Drinking water source and chlorination byproducts. I. Risk of colon and rectal cancers. *Epidemiology*, volume 9, pages 29–35, 1998.

Hoyer, A. P., Grandjean, P., Jorgensen, T., et al. Organochlorine exposure and risk of breast cancer. *Lancet*, volume 352, pages 1816–1820, 1998.

Koivusalo, M., and Vartiainen, T. Drinking water chlorination by-products and cancer. *Reviews of Environmental Health*, volume 12, pages 81–90, 1997.

Koivusalo, M., Pukkala, E., Vartiainen, T., et al. Drinking water chlorination and cancer: A historical cohort study in Finland. *Cancer Causes and Control*, volume 8, pages 192–200, 1997.

Koivusalo, M., Vartiainen, T., Haulinen, T., et al. Drinking water mutagenicity and leukemia, lymphomas, and cancers of the liver, pancreas, and soft tissue. *Archives of Environmental Health*, volume 50, pages 269–276, 1995.

6   *Drinking Water, Estrogens, and Fertility*

A book that extensively reviews the history of how environmental contaminants were found to have a negative impact on the endocrine system is:

Colborn, T., Dumanoski, D., and Myers, J. P. *Our Stolen Future*, Dutton Books, New York, 1996. An extensive bibliography is found in the back of the book.

Other sources of information, with bibliographies, are the following:

Carlsen, E., Giwercman, A., Keiding, N., et al. Evidence of decreasing quality of semen during past 50 years. *British Medical Journal,* volume 305, pages 609–613, 1992.

Clarkson, T. W. Environmental contaminants in the food chain. *American Journal of Clinical Nutrition,* volume 61, supplement 3, pages 682s–686s, 1995.

Danish Environmental Protection Agency. *Male Reproductive Health and Environmental Chemicals with Estrogenic Effects.* Miljoprojekt 290, 1995.

Daston, G. P., Gooch, J. W., Breslin, W. J., et al. Environmental estrogens and reproductive health: A discussion of the human and environmental data. *Reproductive Toxicology,* volume 11, pages 465–481, 1997.

Davis, D. L., Gottlieb, M. B., and Stampnitzky, J. R. Reduced ratio of male to female births in several industrial countries: A sentinel health indicator? *Journal of the American Medical Association,* volume 279, pages 1018–1023, 1998.

Editorial. Male reproductive health and environmental oestrogens. *Lancet,* volume 345, pages 933–935, 1995.

Garcia-Rodriguez, J., Garcia-Martin, M., Nogueras-Ocana, M., et al. Exposure to pesticides and cryptorchidism: Geographical evidence of a possible association. *Environmental Health Perspectives,* volume 104, pages 1090–1095, 1996.

Keith, L. H. *Environmental Endocrine Disruptors: A Handbook of Property Data.* John Wiley and Sons, New York, 1997.

National Institutes of Health. Estrogens in the environment. *Environmental Health Perspectives,* volume 103, supplement 7, pages 1–178. Publication No. NIH 95–218, October 1995.

National Wildlife Federation. Fertility on the brink: The legacy of the chemical age. www.nwf.org/lib/fertility.

Rudel, R. Predicting health effects of exposure to compounds with estrogenic activity: Methodological issues. *Environmental Health Perspectives,* volume 105, supplement 3, pages 655–663, 1997.

Sharpe, R. M., and Skakkebaek, N. E. Are oestrogens involved in falling sperm counts and disorders of the male reproductive tract? *Lancet,* volume 341, pages 1392–1395, 1993.

U.S. Environmental Protection Agency, Office of Prevention, Pesticides, and Toxic Substances. Introduction to the OPPTS Endocrine Disruptors Homepage. www.epa.gov/opptintr/opptendo/endo.

7 *The Effects of Drinking Water's Mineral Content on Health*

Cardiovascular Disease

Bernardi, D., Dini, F. L., Azzarelli, A., et al. Sudden cardiac death in an area characterized by high incidence of coronary artery disease and low hardness of drinking water. *Angiology,* volume 46, pages 145–149, 1995.

Comstock, G. W. Water hardness and cardiovascular diseases. *American Journal of Epidemiology*, volume 110, pages 375–400, 1979.

Editorial. FDA funds Academy of Sciences study on benefits of magnesium. September 30, 1995. www.execpc.com/~magnesium/fdaweek. htlm.

Resnick, L., Bardicef, O., Altura, B., et al. Serum ionized magnesium: Relation to blood pressure and racial factors. *American Journal of Hypertension*, volume 10, pages 1420–1424, 1997.

Rubenowitz, E., Axelsson, G., and Rylander, R. Magnesium in drinking water and death from acute myocardial infarction. *American Journal of Epidemiology*, volume 143, pages 456–462, 1996.

Sharrett, A. R. The role of chemical constituents of drinking water in cardiovascular diseases. *American Journal of Epidemiology*, volume 110, pages 401–419, 1979.

Osteoporosis

Van Dokkum, W., De La Gueronniere, V., Schaafsma, G., et al. Bioavailability of calcium and fresh cheeses, enteral food, and mineral water: A study of stable calcium isotopes in young adult women. *British Journal of Nutrition*, volume 75, pages 893–903, 1996.

Kidney Stones

Churchill, D., Bryant, D., Fodor, G., et al. Drinking water hardness and urolithiasis. *Annals of Internal Medicine*, volume 88, pages 513–514, 1978.

Donaldson, D., Pryce, J. D., Rose, G. A., et al. Tap water calcium and its relationship to renal calculi and 24h urinary calcium output in Great Britain. *Urological Research*, volume 7, pages 273–276, 1979.

Jaeger, P., Portmann, L., Jacquet, A. F., et al. Drinking water for stone formers: Is the calcium content relevant? *European Urology*, volume 10, pages 53–54, 1984.

Seftel, A., and Resnick, M. I. Metabolic evaluation of urolithiasis. *Urologic Clinics of North America*, volume 17, pages 159–169, 1990.

Shuster, J., Finlayson, B., Schaeffer, R., et al. Water hardness and urinary stone disease. *Journal of Urology*, volume 128, pages 422–425, 1982.

Fluoride

Centers for Disease Control and Prevention. Engineering and administrative recommendations for water fluoridation, 1995. *Morbidity and Mortality Weekly Reports*, volume 44 (RR-13), pages 1–40, 1995.

Driscoll, W. S., Horowitz, H. S., Meyers, R. J., et. al. Prevalence of dental caries and dental fluorosis in areas with negligible, optimal, and above-optimal fluoride concentrations in drinking water. *Journal of the American Dental Association*, volume 113, pages 29–33, 1986.

Duckworth, R. M. The science behind caries prevention. *International Dental Journal*, volume 43, supplement 1, pages 529–539, 1993.

Horowitz, H. S., Heifetz, S. B., and Law, F. E. Effect of school fluoridation on dental caries: Final results in Elk Lake, Pennsylvania, after 12 years. *Journal of the American Dental Association*, volume 84, pages 832–838, 1972.

Ismail, A. I. Fluoride supplements: Current effectiveness, side effects, and recommendations. *Community Dentistry and Oral Epidemiology*, volume 22, pages 164–172, 1994.

Katz, R. V., Jensen, M. E., and Meskin, L. H. Dental caries prevalence in a community of schoolchildren: Fluoridated vs. non-fluoridated. *Journal of Dental Research*, volume 62, page 202, 1983.

Newbrun, E., (editor) *Fluorides and Dental Caries*, 3rd edition. C. C. Thomas, Springfield, Illinois, 1986.

History

Coley, N. G. "Cures without care," "chymical physicians," and mineral waters in seventeenth-century English medicine. *Medical History*, volume 23, pages 191–214, 1979.

Coley, N. G. Physicians and the chemical analysis of mineral waters in eighteenth-century England. *Medical History*, volume 26, pages 123–144, 1982.

Weisz, G. Water cures and science: The French academy of medicine and mineral waters in the nineteenth century. *Bulletin of the History of Medicine*, volume 64, pages 393–416, 1990.

8 Heavy Metal Content of Drinking Water and Its Effects on Health

Lead

Eisinger, J. Lead and wine. Eberhard Gockel and the colica Pictonum. *Medical History*, volume 26, pages 279–302, 1982.

Hayes, E. B. The hazards of lead to children. In *Environmental Medicine*. Edited by S. Brooks. Mosby, St. Louis, Illinois, pages 383–389, 1995.

Lockitch, G. Perspectives on lead toxicity. *Clinical Biochemistry*, volume 26, pages 371–381, 1993.

Quinn, M. J., and Sherlock, J. C. The correspondence between U. K. "action levels" for lead in blood and in water. *Food Additives and Contaminants*, volume 7, pages 387–424, 1990.

Rabinowitz, M., Needleman, H., Burley, M., et al. Lead in umbilical blood, indoor air, tap water, and gasoline in Boston. *Archives of Environmental Health*, volume 39, pages 299–301, 1984.

Shannon, M. W., and Graef, J. W. Lead intoxication in infancy. *Pediatrics*, volume 89, pages 87–90, 1992.

Tong, S., Baghurst, P. A., Sawyer, M. G., et al. Declining blood lead levels and changes in cognitive function during childhood. The Port

Pirie Cohort Study. *Journal of the American Medical Association*, volume 280, pages 1915 –1919, 1998.

Waldron, H. A. Lead poisoning in the ancient world. *Medical History*, volume 17, pages 391–399, 1973.

Watt, G. C. M., Briton, A., Gilmour, W. H., et al. Is lead in tap water still a public health problem? An observational study in Glasgow. *British Medical Journal*, volume 313, pages 979–981, 1996.

Copper

Sidhu, K. S., Nash, D. F., and McBride, D. E. Need to revise the national drinking water regulation for copper. *Regulatory Toxicology and Pharmacology*, volume 22, pages 95–100, 1995.

Aluminum

Doll, R. Alzheimer's disease and environmental aluminum. *Age and Ageing*, volume 22, pages 138–153, 1993.

Editorial. Is aluminum a dementing ion? *Lancet*, volume 339, pages 713–714, 1992.

Martyn, C. N., Osmond, C., Edwardson, J. A., et al. Geographical relation between Alzheimer's disease and aluminum in drinking water. *Lancet*, volume 1, pages 59–62, 1989.

McLachlan, D. R., Bergero, C., Smith, J. E., et al. Risk of neuropathologically confirmed Alzheimer's disease and residual aluminum in municipal drinking water employing weighted residential histories. *Neurology*, volume 46, pages 401–405, 1996.

Mercury

Clarkson, T. W. Mercury: Major issues in environmental health. *Environmental Health Perspectives*, volume 100, pages 31–38, 1992.

Davidson, P. W., Myers, G. J., Cox, C., et al. Effects of prenatal and postnatal methylmercury exposure from fish consumption on neurodevelopment: Outcomes at 66 months in the Seychelles Child Development Study. *Journal of the American Medical Association*, volume 280, pages 701–707, 1998.

Grandjean, P., Weihe, P., and Nielsen, J. B. Methylmercury: Significance of intrauterine and postnatal exposures. *Clinical Chemistry*, volume 40, pages 1395–1400, 1994.

Grandjean, P., Weihe, P., Jorgensen, P. J., et al. Impact of maternal seafood diet on fetal exposure to mercury, selenium, and lead. *Archives of Environmental Health*, volume 47, pages 185–195, 1992.

Grandjean, P., Weihe, P., and White, R. F. Milestone development in infants exposed to methylmercury from human milk. *Neurotoxicology*, volume 16, pages 27–33, 1995.

Greenwood, M. R. Methylmercury poisoning in Iraq: An epidemiological study of the 1971–1972 outbreak. *Journal of Applied Toxicology*, volume 5, pages 148–159, 1985.

Lindberg, S. E. Emission and deposition of atmospheric mercury vapor. In: *Lead, Mercury, Cadmium, and Arsenic in the Environment.* Edited by T. C. Hutchinson and K. M. Meena. John Wiley and Sons, New York, pages 89–106, 1987.

Nirenberg, D. W., Nordgren, R. E., Chang, M. B., et al. Delayed cerebellar disease and death after accidental exposure to dimethylmercury. *New England Journal of Medicine,* volume 338, pages 1672–1677, 1998.

9 Bottled Water

Two books that the reader can refer to regarding bottled water were written by Arthur von Weisenberger:

von Weisenberger, A. *H₂O*. Woodbridge Press, Santa Barbara, California 1988.

von Weisenberger, A. *The Pocketbook to Bottled Water.* Contemporary Books, Chicago, Illinois, 1991.

Other books available on drinking water are:

Bottled Water Regulation. Hearing before the Subcommittee on Oversight and Investigations of the Committee on Energy and Commerce, House of Representatives, 102nd Congress, First Session, April 10, 1991. Available through www.amazon.com.

Green, M., and Green, T. *Good Water Guide: The World's Best Bottled Water.* Seven Hills Book distributors, 1996. Available through www.amazon.com.

Green, M., and Green, T. *The Best Bottled Water in the World.* 1986. Available through www.amazon.com.

Two web sites with information about bottled water are:

www.bottledwaterweb.com. Information on various bottling companies is available at this site.

www.bottledwater.org. Information from the International Bottled Water Association is presented at this site.

10 Water Purification

The best place to look up information regarding water purification and water purifiers is on the web site of the National Sanitation Foundation, www.nsf.org. Here one can look up sites of individual manufacturers and suppliers. The organization's mailing address is National Sanitation Foundation, 3475 Plymouth Road, P.O. 1468, Ann Arbor, MI 48106 (telephone: 800 673-8010).

Another source of information is the Water Quality Association, Consumer Affairs Department, P.O. 606, Lisle, IL 60532 (telephone: 800 749-0234; web site: www.wqa.com).

Other sources are:

Ogerth, J. E., Johnson, R. L., MacDonald, S. C., et al. Backcountry water treatment to prevent giardiasis. *American Journal of Public Health,* volume 79, pages 1633–1637, 1989.

Towles, J., and Edwards, H. B. The use of ozone in the purification of swimming pool water. www.ozonated.com.

Index

About the Authors

Joshua I. Barzilay is an endocrinologist with the Southeast Permanente Medical Group in Atlanta, Georgia. **Winkler G. Weinberg** is an infectious disease specialist with the Southeast Permanente Medical Group. Dr. Weinberg is the author of *No Germs Allowed!*, a book on infectious diseases published by Rutgers Universtiy Press. **J. William Eley** is an oncologist and epidemiologist at the Emory University School of Medicine and the Rollins School of Public Health at Emory University.